Network Communications Technology

Network Communications Technology

Ata Elahi, Ph.D

Southern Connecticut State University
Computer Science Department

Delmar
Thomson Learning™

Africa • Australia • Canada • Denmark • Japan • Mexico
New Zealand • Philippines • Puerto Rico • Singapore
Spain • United Kingdom • United States

Delmar Staff:

Business Unit Director: Alar Elken
Executive Editor: Sandy Clark
Acquisitions Editor: Gregory L. Clayton
Developmental Editor: Michelle Ruelos Cannistraci
Editorial Assistant: Jennifer Thompson
Executive Marketing Manager: Maura Theriault
Channel Manager: Mona Caron

Marketing Coordinator: Paula Collins
Executive Production Manager: Mary Ellen Black
Production Manager: Larry Main
Senior Project Editor: Christopher Chien
Art and Design Coordinator: David Arsenault

For more information, contact Delmar, 3 Columbia Circle, PO Box 15015, Albany, NY 12212-0515; or find us on the World Wide Web at http://www.delmar.com.

Asia
Thomson Learning
60 Albert Street, #15-01
Albert Complex
Singapore 189969

Australia/New Zealand
Nelson/Thomson Learning
102 Dodds Street
South Melbourne, Victoria 3205
Australia

Canada
Nelson/Thomson Learning
1120 Birchmont Road
Scarborough, Ontario
Canada M1K 5G4

International Headquarters
Thomson Learning
International Division
290 Harbor Drive, 2nd Floor
Stamford, CT 06902-7477
USA

Japan
Thomson Learning
Palaceside Building 5F
1-1-1 Hitotsubashi, Chiyoda-ku
Tokyo 100 0003 Japan

Latin America
Thomson Learning
Seneca, 53
Colonia Polanco
11560 Mexico D.F. Mexico

Spain
Thomson Learning
Calle Magallanes, 25
28015-Madrid
Espana

UK/Europe/Middle East
Thomson Learning
Berkshire House
168-173 High Holborn
London
WC1V 7AA United Kingdom

Thomas Nelson & Sons Ltd.
Nelson House
Mayfield Road
Walton-on-Thames
KT 12 5PL United Kingdom

ISBN: 07668-1388-6

Contents

Preface

This book is a result of my teaching the course, Data Communications and Computer Networks at Southern Connecticut State University since 1987. The book covers the technology aspect of networks, rather than the theories of networks. The beta version of this textbook was tested in undergraduate level computer network courses at Southern Connecticut State University. The textbook covers networking using a *direct, practical approach* that explains the technology in simple terms. This book covers the latest topics in networking technology, such as Digital Subscriber Line (DSL), cable modems, asynchronous transfer mode (ATM), Fast Ethernet, LAN switching, and Gigabit Ethernet.

Intended Audience

This book is written primarily as an introduction to networking for students majoring in computer science technology, electronics technology, or engineering. The readability is clear and easy to understand for those in technical colleges, while the broad range of topics is appealing for those in higher level courses. For students in business application and CIS courses, the instructor may wish to omit parts of the text.

Organization

The material in this textbook is presented *practically* rather than taking a theoretical and mathematical approach. Therefore, no specialized background is required to understand the material; the first three chapters of this book form the foundation for the rest of the text. I have opted to focus on the technology aspect of networks, so each technology is presented in a separate chapter. In addition, I offer a brief introduction to computer architecture, because networks are simply another part of the computer.

Chapter 1 is an introduction to computer networks, network topologies, and types of networks.

Chapter 2 covers basic data communications, including digital signals, binary numbers, serial and parallel transmission, communication modes, digital encoding, and error detection methods used in networking. This chapter gives students the basic knowledge required for the rest of the textbook.

Chapter 3 presents an overview of computer architecture. The network is part of the computer; therefore, the reader should have a basic knowledge of computer architecture. This chapter covers the basic components of a microcomputer: the CPU, types of memory, and computer buses.

Chapter 4 covers data communication media such as twisted-pair cable, coaxial cable, fiber-optic cable, and wireless communication.

Chapter 5 covers the multiplexer, demultiplexer, types of multiplexers, T1 link architecture, and switching concepts.

Chapter 6 presents modem technology, modulation methods, Digital Subscriber Line (DSL) technology, and cable modem technology.

Chapter 7 explains the function of standards organizations and lists some computer protocols followed by a summary of the Open System Model, which explains the function of each layer. IEEE 802 committee standards are presented.

Chapter 8 covers the Ethernet Network, from operation to technical specifications and cabling. It serves as the foundation for the material on Fast Ethernet and Gigabit Ethernet covered in Chapters 10 and 13.

Chapter 9 explains the operation of a Token Ring Network, token ring technical specifications, and token bus operation.

Chapter 10 presents Fast Ethernet technology, Fast Ethernet repeaters, and different types of media used for Fast Ethernet.

Chapter 11 covers 100VG–AnyLAN networking technology and specifications.

Chapter 12 presents switching technology and its applications. It also covers VLAN operation and Firewall technology.

Chapter 13 explains Gigabit Ethernet technology and the types of media used for Gigabit Ethernet, followed by an explanation of the applications of Gigabit Ethernet.

Chapter 14 presents networking interconnection devices such as repeaters, bridges, routers, and gateways.

Chapter 15 covers Fiber Distributed Data Interface (FDDI) technology and its applications.

Chapter 16   presents SONET components and architecture.

Chapter 17   explains Narrowband ISDN and its applications.

Chapter 18   covers Frame Relay technology applications and components.

Chapter 19   explains Internet Architecture, Transmission Control Protocol and Internet Protocol, IPv6, and Internet II.

Chapter 20   covers the application of an ATM network, ATM network components, ATM switch architecture, and ATM adaptation layers.

Chapter 21   provides a brief overview of Networking Operating Systems, such as Windows NT and Novell NetWare.

Acknowledgments

I would like to express my special thanks to Mr. Mike Rizzo, Dr. Lisa Lancor, Professor Winnie Yu, and Mr. Richard Pair for their guidance and suggestions on how to improve this textbook. Also, I would like to thank the following reviewers:

Mike Awwad, DeVry Institute of Technology, North Brunswick, NJ

Omar Ba-Rukab, Florida Institute of Technology, Melbourne, FL

Robert Diffenderfer, DeVry Institute of Technology, Kansas City, MO

Enrique Garcia, Laredo Community College, Laredo, TX

Dr. Rafiqul Islam, DeVry Institute of Technology, Alberta, CANADA

Lynnette Garetz, Heald College, Hayward, CA

Anu Gokhale, Illinois State University, Normal, IL

Nebojsa Jansic, DeVry Institute of Technology, Columbus, OH

Steve Kuchler, IVY Tech, Indianapolis, IN

William Lin, Indiana University–Purdue University, Indianapolis, IN

Clifford Present, DeVry Institute of Technology, Pomona, CA

Glenn Rettig, Owens Community College, Findlay, OH

Richard Rouse, DeVry Institute of Technology, Kansas City, MO

Mohammed Mehdi Shahsavari, Florida Institute of Technology, Melbourne, FL

Lowell Tawney, DeVry Institute of Technology, Kansas City, MO

Julius Willis, Heald College, Hayward, CA

chapter 1

Introduction to Computer Networks

OBJECTIVES

After completing this chapter, you should be able to:

- Explain the advantages of computer networks
- Describe the components of a network
- Discuss the function of a client/server model
- Explain various networking topologies
- Describe different types of networks in terms of advantages and disadvantages

INTRODUCTION

Networking is a business tool for companies: for example, a bank can transfer funds between branches by using a network. People can access their bank accounts using automatic teller machines via a network. Travel agencies are using networks to make airline reservations. We can even make banking transactions and shop on-line using a network. Networking technology is growing fast and will enable students to access the Internet in any location on their campus by using their laptop computers.

Network is a generic term. Several computers connected together are called a **computer network**. A network is a system of computers and related equipment connected by communication links to share data. The related equipment may be printers, fax machines, modems, copiers, and so forth. Following are some of the benefits of using computer networks:

Resource Sharing: Computers in a network can share resources such as data, printers, disk drives, scanners, and so on.

Reliability: Since computers in a network can share data, if one of the computers on the network crashes, a copy of its resources might be found on other computers in the network.

1

Cost: Microcomputers are much less expensive than mainframes. Instead of using several mainframes, a network can use one mainframe as a server, with several microcomputers connected to the server as clients. This creates a client/server relationship.

Communication: Users can exchange messages via electronic mail or other messaging systems, or they can transfer files.

1.1 Network Models

A computer in a network can be either a server or a client. A **server** is a computer on the network that holds shared files and the network operating system that manages the network operation. In order for a **client** computer to access the resources in the server, the client computer must request information from the server. The server will then transmit the information requested to the client.

Three models are used, based on type of network operation needed. They are:

1. Peer-to-Peer network (Work Group)
2. Server-Based network
3. Client/Server network

Peer-to-Peer Model (Work Group)

In a peer-to-peer network there is no special station that holds shared files and a network operating system. Each station can access the resources of other stations in the network. Individual stations can act as server and/or client. In this model, each user is responsible for administrating and upgrading the software of his or her station. Since there is no centralized station to manage network operation, this model is used for a network of fewer than ten stations. Figure 1.1 shows peer-to-peer network model.

FIGURE 1.1
Peer-to-peer
network

Server-Based Model

In a server-based network, a server stores all the network's shared files and applications such as word processor documents, compilers, database applications, and the network operating system (NOS) that manages network operations. A user can access the file server and transfer shared files to his or her station.

Figure 1.2 shows a network with one file server and three users, or clients. Each client can access the resources on the server, and also the resources of other clients. Clients that are connected on a network are able to freely exchange information with one another. Some of the most popular servers are:

FIGURE 1.2
Network with one
server and three
clients

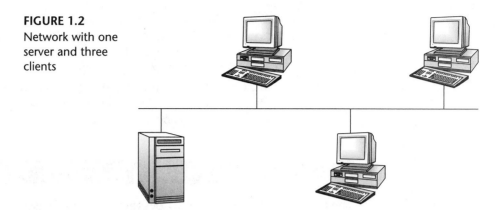

Mail server: A mail server stores all the clients' mail. The client can access the server and transfer incoming mail to its station. Clients can also use the mail server to transfer outgoing mail to the networks.

Print server: Client can submit files to a print server for printing.

Communication server: The server is used by clients to communicate with other networks via modem or other communication link.

Client/Server Model

In the **client/server model,** a client submits its task to the server; the server then executes the client's task and returns the results to the requesting client station. This method of information sharing is called the client/server model and is depicted in Figure 1.3. In a client/server model, less information travels through the network compared to the file server model, making more efficient use of the network.

FIGURE 1.3
Client/Server
model

Server   Clients

1.2   Network Components

A network is composed of several components. The basic components of a computer network are listed below.

1. **Network Interface Card (NIC):** Each computer in a network requires a Network Interface Card. The NIC allows the stations on the network to communicate with each other.

2. **Transmission Medium:** The transmission medium connects the computers together and provides a communication link between the computers on the network. Some of the more common types are twisted-pair cable, coaxial cable, and fiber-optic cable.

3. **Network Operating System (NOS):** The NOS runs on the server and provides services to the client such as login, password, print file, network administration, and file sharing.

1.3 Network Topology

The **topology** of a network describes the way computers are connected together. Topology is a major design consideration for cost and reliability. Following is a list of common topologies found in computer networking:

- Star
- Ring
- Bus
- Fully-Connected Network or Mesh
- Tree
- Hybrid

Star Topology In a **star topology**, all stations are connected to a central controller or hub as shown in Figure 1.4. For any station to communicate with another station, the source must send information to the hub, then the hub must transmit that information to the destination station. If station #1 wants to send information to station #3, it must send information to the hub; the hub must then pass the information to station #3.

FIGURE 1.4
Star topology

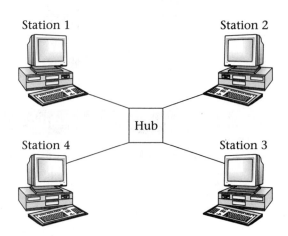

The disadvantage of star topology is that the operation of the entire network depends on the hub. If the hub breaks down, the entire network is disabled. The advantages of star topology are as follows:

- It is easy to set up the network.
- It is easy to expand the network.
- If one link to the hub breaks, only the station using that link is affected.

Ring Topology IBM invented ring topology, which is well known as IBM Token Ring. In a **ring topology** all the stations are connected in cascading order to make a ring, as shown in Figure 1.5. The source station transfers information to the next station on the ring, which checks the address of the information. If the address of the information matches with the station's address, the station copies the information and passes it to the next station; the next station repeats the process and passes the information on to the next station, and so on, until the information reaches the source station. The source then removes the information from the ring. The arrows in Figure 1.5 indicate the direction in which the information flows.

FIGURE 1.5
Ring topology

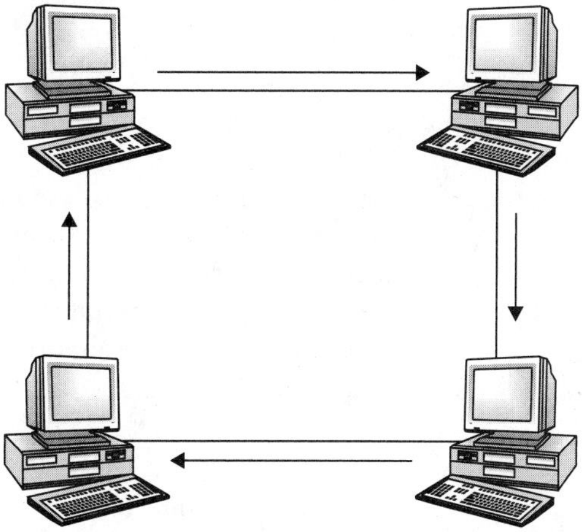

The disadvantages of ring topology are as follows:

- If a link or a station breaks down, the entire network is disabled.
- Complex hardware is required (the network interface card is expensive).
- Adding a new client disrupts the entire network.

The advantages of ring topology are as follows:

- It is easy to install.
- It is east to expand.
- It can use fiber-optic cable.

Bus Topology   A **bus network** is a multi-point connection in which stations are connected to a single cable called a bus. Bus topology is depicted in Figure 1.6. In bus topology, all stations share one media. Ethernet is one of the most popular LANs that uses bus topology. The bus topology is one of the most popular topologies used in LAN networking.

The advantages of bus topology are simplicity, low cost, and easy expansion of the network. The disadvantage of bus topology is that a breakdown in the bus cable brings the entire network down.

FIGURE 1.6
Bus topology

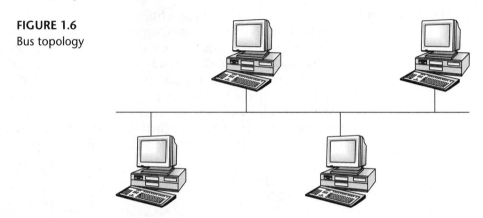

Fully-Connected   A **fully-connected topology** connects each station directly to every other sta-
Topology   tion in the network, as shown in Figure 1.7.

The advantage of a fully-connected topology (mesh topology) is that each station has a dedicated connection to the other stations, therefore this topology offers the highest reliability and security. If one link in the mesh topology breaks, the network remains active.

FIGURE 1.7
Fully-connected
topology

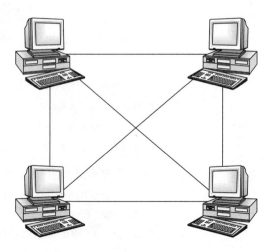

A major disadvantage of a fully-connected topology is that it uses too many connections and therefore requires a great deal of wiring, especially when the number of stations increases. Consider, for example, a fully-connected network with 100 workstations. Workstation #1 would require 99 network connections to connect it to workstations 2 through 99. The number of connections is determined by $N(N - 1)/2$, where N is the number of stations in the network. This type of topology is not cost effective.

Tree Topology **Tree topology** uses an active hub or repeater to connect stations together. The **hub** is one of the most important elements of a network because it links all stations in the network together. The function of the hub is to accept information from one station and repeat the information to other stations and hub, as shown in Figure 1.8.

FIGURE 1.8
Tree topology

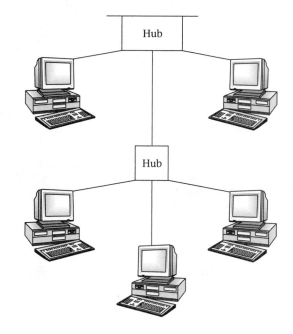

The advantage of this topology is that when one hub breaks, only stations connected to the broken hub will be affected. There are several types of hubs listed below.

Manageable Hub: Intelligent hubs are defined as manageable hubs, which means each of the ports on the hub can be enabled or disabled by the network administrator through software.

Stand-Alone Hub: A stand-alone hub is a type of hub used for workgroups of computers that are separate from the rest of the network. They cannot be linked together logically to represent a larger hub.

Modular Hub: A modular hub comes with a chassis or card cage and the number of ports can be extended by adding extra cards.

Stackable Hub: A stackable hub looks like a stand-alone hub, but several of them can be stacked or connected together in order to increase the number of ports.

Hybrid Topology   **Hybrid topology** is a combination of different topologies connected together by a backbone cable as shown in Figure 1.9. Each network is connected to the backbone cable by a device called a bridge.

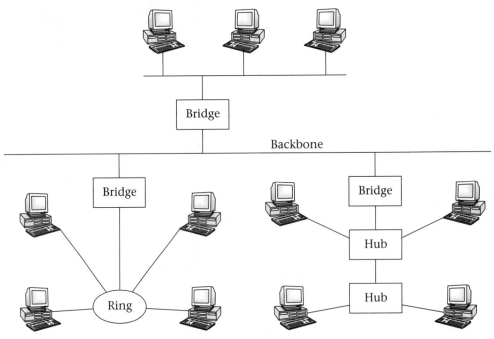

FIGURE 1.9   Hybrid topology

1.4   Types of Networks

The distance between computers that are connected as a network determines the type of network, such as Local Area Network (LAN), Metropolitan Area Network (MAN), and Wide Area Network (WAN).

Local Area Network (LAN)   A **Local Area Network (LAN)** is a high-speed network designed to link computers and other data communication systems together within a small geographic area such as an office, department, or a single floor of a multistory building.

Several LANs can be connected together in a building or on a campus to extend the connectivity. A LAN is considered a private network. The most popular LANs in use today are Ethernet, Token Ring, and Gigabit Ethernet.

Metropolitan Area Network (MAN)   **Metropolitan Area Networks (MAN)** cover approximately 100 miles, connecting multiple networks which are located in different locations of a city or town. The communication links in a MAN are generally owned by a network service provider. Figure 1.10 shows a Metropolitan Area Network.

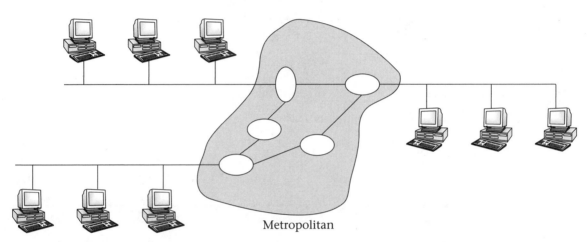

Metropolitan

FIGURE 1.10   Metropolitan area network

Wide Area Networks (WANs)   A **Wide Area Network (WAN)** is used for long-distance transmission of information. WANs cover a large geographical area, such as an entire country or continent. WANs may use lines leased from telephone companies, or Public Switched Data Networks (PSDN), or satellites for communication links.

Internet   The **Internet** is a collection of networks located all around the world that are connected by gateways, as shown in Figure 1.11. Each gateway has a routing table containing information about the networks that are connected to the gateway. Several networks may be connected to one gateway. The gateway accepts information from the network and checks its routing table to see if the destination is in the network connected to the gateway. If the destination station is in the network connected to the gateway, it transmits the information to the network. Otherwise it passes the information to the next gateway, which performs the same operation; this process continues until the information reaches its destination.

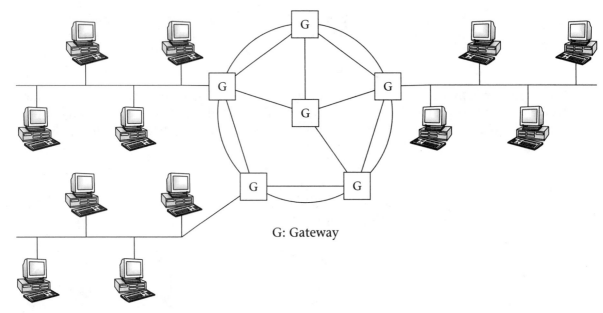

FIGURE 1.11   Internet architecture

Summary

- A group of several computers connected by communication media is termed a **computer network**.
- Some of the application of computer networks are file sharing, hardware sharing, and electronic mail.
- In the **client/server** model of networking, the client submits the information to the server, which processes the information and returns the results to the client station.
- The components of a network are: Network Interface Card (NIC), Network Operating System (NOS), and the communication link (transmission medium).
- Computers can be connected in the form of **Star**, **Ring**, **Bus**, **Fully-Connected**, **Tree** and **Hybrid Topologies**.
- The types of networks are: **LAN**, **MAN**, **WAN**, and **Internet**.

Key Terms

Bus Network	Client/Server Model
Client	Computer Network

Fully-Connected Topology   Ring Topology

Hybrid Topology   Server

Internet   Star Topology

Local Area Network (LAN)   Topology

Metropolitan Area Network (MAN)   Transmission Medium

Network Interface Card (NIC)   Tree Topology

Network Operating System (NOS)   Wide Area Network (WAN)

Review Questions

● **Multiple Choice Questions**

1. Several computers connected together are called a _____.
   - a. computer network
   - b. client
   - c. server
   - d. hub

2. In a _____ network, the client submits a task to the server, then the server executes and returns the result to the requesting client station.
   - a. Peer-to-Peer
   - b. Server-Based
   - c. Client/Server
   - d. all of the above

3. A computer in a network can function as a _____ or _____.
   - a. client, server
   - b. client, user
   - c. a and b
   - d. none of the above

4. A _____ stores all the client's mail.
   - a. file server
   - b. print server
   - c. communication server
   - d. mail server

5. A _____ uses a modem or other type of communication link to enable clients to communicate with other networks.
   - a. mail server
   - b. communication server
   - c. a and b
   - d. none of the above

6. In a _____ topology, all stations are connected to a central controller or hub.
   - a. star
   - b. ring
   - c. bus
   - d. fully-connected

7. In a _____ topology, all stations are connected in cascade.
   - a. star
   - b. ring
   - c. tree
   - d. bus

8. A _____ topology is a combination of different topologies connected together by a backbone cable.
   - a. star
   - b. ring
   - c. bus
   - d. hybrid

9. Which network topology uses a hub? _____
   - a. Ring
   - b. Bus
   - c. Star
   - d. Mesh

10. Which type of topology uses multi-point connections? _____
    - a. Bus
    - b. Star
    - c. Ring
    - d. Fully-connected

11. How many connections are required by a fully-connected network with five stations? _____
    - a. 5
    - b. 10
    - c. 20
    - d. 15

12. Which of the following networks is used for office buildings? _____
    - a. LAN
    - b. MAN
    - c. WAN
    - d. Internet

13. Which of the following topologies is used for Ethernet? _____
    - a. Bus
    - b. Star
    - c. Ring
    - d. Mesh

14. The Internet is a collection of LANs connected together by _____.
    - a. routers
    - b. switches
    - c. gateways
    - d. repeaters

15. Computers on a campus are connected by a/an _____.
    - a. LAN
    - b. WAN
    - c. MAN
    - d. Internet

● **Short Answer Questions**

1. Explain the function of a server.
2. What is the function of the client in a file server model?
3. Explain the term client/server model.
4. What are the advantages of client/server model?
5. What are the network components?
6. A Network Operating System runs on _____.
7. List the networking topologies.
8. What is the disadvantage of a fully-connected topology?
9. What is a hub?
10. What are the types of hubs?
11. What does MAN stand for?
12. Explain Internet operation.
13. What does WAN stand for?
14. What are advantages of bus topology?

chapter 2

Introduction to Data Communication

OBJECTIVES   After completing this chapter, you should be able to:

- Distinguish between analog and digital signals
- Convert decimal numbers to binary, binary to hexadecimal, and vice versa
- Represent characters and decimal numbers in 7-bit ASCII code
- Compare serial, parallel, asynchronous, and synchronous transmission
- List the communication modes
- Explain the different types of digital encoding methods
- Calculate Block Check Character (BCC) and Frame Check Sequence (FCS), which are used for error detection in networking

INTRODUCTION   In order to understand network technology it is important to know how information is represented for transmission from one computer to another. Information can be transferred between computers in one of two ways: by using either an analog signal or a digital signal. Figure 2.1 shows the analog signal of a sine wave that is repeated every "T" seconds. T is the period of the signal, and it represents the time of one cycle.

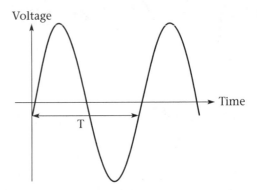

FIGURE 2.1
Analog signal

2.1 Characteristics of Analog Signals

Computers can be built in two forms: digital and analog. The principle feature of analog representation is that the signals are continuous. Watches with hands are analog because the hands move continuously around the clock face. Digital watches are called digital because they move from one value to the next without displaying all of the intermediate values.

An analog signal has continuous value and can have any value at any time. A periodic analog signals varies regularly between the values over time and can be represented graphically as a wave. The simplest form of analog signal is the sine wave, as shown in Figure 2.1. The characteristics of an analog signal are frequency, amplitude, and phase.

Frequency   Frequency (F) is the number of cycles in one second or F = 1/T, represented in Hertz (Hz). If the cycle of an analog signal is repeated every one second, the frequency of the signal is one hertz. If the cycle of an analog signal is repeated 1000 times every second (once every millisecond) the frequency is:

$$F = 1/T = 1/10^{-3} = 1000 \text{ Hz} = 1 \text{ kHz}$$

Note: 1,000 Hz is equal to 1 kilohertz (kHz) and 1,000,000 Hz is equal to 1 megahertz (MHz).

Amplitude   The amplitude of an **analog signal** is a function of time (as shown in Figure 2.2) and can be represented in volts (unit of voltage). In other words, the amplitude of an analog signal is its voltage value at any given time. At the time of t_1, the amplitude of signal is V_1.

Phase   Two signals with the same frequency can differ in phase. This means that one of the signals starts at a different time from the other signal. This difference can be represented by degrees, from 0 to 360 degrees. Figure 2.3 shows signal B shifted 90 degrees and signal C shifted 270 degrees.

FIGURE 2.2
Amplitude of an
analog signal

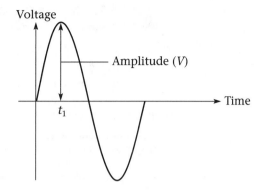

FIGURE 2.3
Phases of an
analog signal

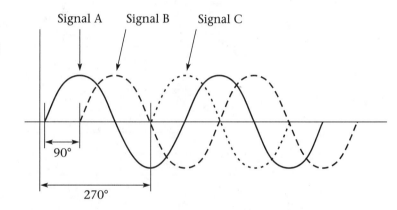

2.2   Digital Signals

Modern computers communicate by using digital signals. **Digital signals** are represented by two voltages: one voltage represents the number 0 in binary and the other voltage represents the number 1 in binary. An example of a digital signal is shown in Figure 2.4, where 0 volts represents 0 in binary and +5 volts represents 1 in binary.

FIGURE 2.4
Digital signal

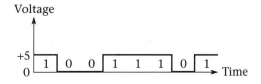

2.3 Binary Numbers

Binary, or Base-2 numbers, are represented by 0 and 1. A binary digit, 0 or 1, is called a **bit**. Eight bits are equal to one **byte**. Two or more bytes are called a **word**. The hexadecimal number system has a base of 16, and therefore has 16 symbols (0 through 9, and A through F). Table 2.1 shows the decimal numbers, their binary values from 0 to 15, and their hexadecimal equivalents.

TABLE 2.1 Decimal Numbers with Binary and Hexadecimal Equivalents

Decimal	Binary (base 2)	Hexadecimal (base 16) or HEX
0	0000	0
1	0001	1
2	0010	2
3	0011	3
4	0100	4
5	0101	5
6	0110	6
7	0111	7
8	1000	8
9	1001	9
10	1010	A
11	1011	B
12	1100	C
13	1101	D
14	1110	E
15	1111	F

Converting to Binary Table 2.1 can also be used to convert a number from hexadecimal to binary and from binary to hexadecimal.

Example 2.1

Convert the binary number 001010011010 to hexadecimal. Each four bits are grouped from right to left. By using Table 2.1, each 4-bit group can be converted to its hexadecimal equivalent.

0010	1001	1010
2	9	A

Example 2.2

Convert $(3D5)_{16}$ to binary. By using Table 2.1, the result in binary is

3	D	5
0011	1101	0101

Example 2.3

Convert 6DB from hexadecimal to binary. By using Table 2.1, the result in binary is

6	D	B
0110	1101	1011

The resulting binary number is: 011011011011

Converting from Binary to Decimal

In general, any binary number can be represented by Equation 2.1.

$$(a_5a_4a_3a_2a_1a_0 .a_{-1}a_{-2}a_{-3})_2 \qquad \text{(Equation 2.1)}$$

where

a_i is a binary digit (either 0 or 1)

Equation 2.1 can be converted to decimal number by using Equation 2.2.

$$(\underbrace{a_5a_4a_3a_2a_1a_0}_{\text{Integer}} .\underbrace{a_{-1}a_{-2}a_{-3}}_{\text{Fraction}})_2 = a_0 \times 2^0 + a_1 \times 2^1 + a_2 \times 2^2 + a_3 \times 2^3 \text{ (Equation 2.2)}$$
$$+ \cdots + a_{-1} \times 2^{-1} + a_{-2} \times 2^{-2} + \cdots$$

Example 2.4

To convert $(110111.101)_2$ to decimal:

$$(110111.101)_2 = 1*2^0 + 1*2^1 + 1*2^2 + 0*2^3 + 1*2^4 + 1*2^5 + 1*2^{-1} + 0*2^{-2}$$
$$+ 1*2^{-3} = 55.625$$

2.4 Coding Schemes

Since computers can only understand binary numbers (0 or 1), all information sent to the computer (such as numbers, letters, and symbols) must be represented as binary data. One commonly used code to represent printable and nonprintable characters is the American Standard Code for Information Interchange (ASCII).

ASCII Code   Each character in ASCII code is represented by seven bits. Table 2.2 shows the **ASCII code** and its hexadecimal equivalent.

Characters from hexadecimal 00 to 1F are control characters, which are nonprintable characters, such as NUL, SOH, STX, ETX , ESC and DLE (data link escape).

Example 2.5 ————————————————————————————————

Convert the word "Network" to binary and show the result in hexadecimal. By using Table 2.2 each character is represented by seven bits and results in:

N	e	t	w	o	r	k
1001110	1100101	1110100	1110111	1101111	1110010	1101011

or in hexadecimal

4E	65	74	77	6F	72	6B

Universal Code (Unicode)   Unicode is a new 16-bit character encoding standard for representing characters and numbers in most languages such as Latin, Greek, Arabic, Chinese, and Japanese. The ASCII code uses eight bits to represent each character in Latin, and it can represent 256 characters. ASCII code does not support mathematical symbols and scientific symbols. **Unicode** uses sixteen bits, which can represent 64,000 characters and symbols. A character in Unicode is represented by16-bit binary, equivalent to four digits in hexadecimal. For example, the character B in Unicode is U0042H (U represents Unicode). The ASCII code is represented from $(00)_{16}$ through $(FF)_{16}$. For converting ASCII code to Unicode, two zeros will be added to the left side of ASCII code; therefore, the Unicode to represent ASCII characters is from $(0000)_{16}$ through $(00FF)_{16}$. For example, the character A is represented by U0041H. Table 2.3 shows a sample of Unicode for Latin and Greek characters. Unicode is divided into blocks of code, with each block assigned to a specific language. Table 2.4 shows each block of Unicode for several languages.

2.5 Transmission Modes

When data is transferred from one computer to another by digital signals, the receiving computer must distinguish the size of each signal to determine when one signal ends and when the next one begins. For example, when a computer sends a signal as shown in Figure 2.5, the receiving computer has to recognize how many ones and zeros are in the signal. Synchronization methods between source and destination devices are generally grouped into two categories: asynchronous and synchronous.

TABLE 2.2 American Standard Code for Information Interchange (ASCII)

Binary	Hex	Char	Binary	Hex	Char	Binary	Hex	Char	Binary	Hex	Char
0000000	00	NUL	0100000	20	SP	1000000	40	@@	1100000	60	'
0000001	01	SOH	0100001	21	!	1000001	41	A	1100001	61	a
0000010	02	STX	0100010	22	"	1000010	42	B	1100010	62	b
0000011	03	ETX	0100011	23	#	1000011	43	C	1100011	63	c
0000100	04	EOT	0100100	24	$	1000100	44	D	1100100	64	d
0000101	05	ENQ	0100101	25	%	1000101	45	E	1100101	65	e
0000110	06	ACK	0100110	26	&	1000110	46	F	1100110	66	f
0000111	07	BEL	0100111	27	'	1000111	47	G	1100111	67	g
0001000	08	BS	0101000	28	(	1001000	48	H	1101000	68	h
0001001	09	HT	0101001	29	)	1001001	49	I	1101001	69	i
0001010	0A	LF	0101010	2A	*	1001010	4A	J	1101010	6A	j
0001011	0B	VT	0101011	2B	+	1001011	4B	K	1101011	6B	k
0001100	0C	FF	0101100	2C	,	1001100	4C	L	1101100	6C	l
0001101	0D	CR	0101101	2D	–	1001101	4D	M	1101101	6D	m
0001110	0E	SO	0101110	2E	.	1001110	4E	N	1101110	6E	n
0001111	0F	SI	0101111	2F	/	1001111	4F	O	1101111	6F	o
0010000	10	DLE	0110000	30	0	1010000	50	P	1110000	70	p
0010001	11	DC1	0110001	31	1	1010001	51	Q	1110001	71	q
0010010	12	DC2	0110010	32	2	1010010	52	R	1110010	72	r
0010011	13	DC3	0110011	33	3	1010011	53	S	1110011	73	s
0010100	14	DC4	0110100	34	4	1010100	54	T	1110100	74	t
0010101	15	NACK	0110101	35	5	1010101	55	U	1110101	75	u
0010110	16	SYN	0110110	36	6	1010110	56	V	1110110	76	v
0010111	17	ETB	0110111	37	7	1010111	57	W	1110111	77	w
0011000	18	CAN	0111000	38	8	1011000	58	X	1111000	78	x
0011001	19	EM	0111001	39	9	1011001	59	Y	1111001	79	y
0011010	1A	SUB	0111010	3A	:	1011010	5A	Z	1111010	7A	z
0011011	1B	ESC	0111011	3B	;	1011011	5B	[	1111011	7B	[
0011100	1C	FS	0111100	3C	<	1011100	5C	\	1111100	7C	\
0011101	1D	GS	0111101	3D	=	1011101	5D	]	1111101	7D	}
0011110	1E	RS	0111110	3E	<	1011110	5E	^	1111110	7E	~
0011111	1F	US	0111111	3F	?	1011111	5F	_	1111111	7F	DEL

TABLE 2.3   Unicode Values for Some Latin and Greek Characters

Latin		Greek	
Character	Code (Hex)	Character	Code (Hex)
A	U0041	φ	U700F
B	U0042	α	U03B1
C	U0043	γ	U03B3
0	U0030	μ	U400B
8	U0038	β	UF00D

TABLE 2.4   Unicode Block Allocations

Start Code (Hex)	End Code (Hex)	Block Name
U0000	U007F	Basic Latin
U0008	U00FF	Latin Supplement
U0037	U03FF	Greek
U0530	U058F	Armenian
U0590	U05FF	Hebrew
U0600	U06FF	Arabic
U01A0	U10FF	Georgian

FIGURE 2.5
Digital signal

Asynchronous Transmission

Asynchronous transmission occurs character by character and is used for serial communication, such as by modem or serial printer. In asynchronous transmission each data character has a start bit, which identifies the start of the character, and one or two bits which identify the end of the character, as shown in Figure 2.6. The data character may consist of five, six, seven, or eight bits. Following the data bits is a parity bit, which is used by the receiver for error detection. After the parity bit is sent, the signal must return to high for at least one bit time to identify the end of the character. The new start bit serves as an indicator to the receiving device that a data character is coming and allows the receiving side to synchronize its clock. Since receiver and transmitter clocks are not synchronized continuously, the transmitter uses the start bit to reset the receiver clock so that it matches the transmitter clock. Also, the receiver is already programmed to recognize the number of bits in each character sent by the transmitter.

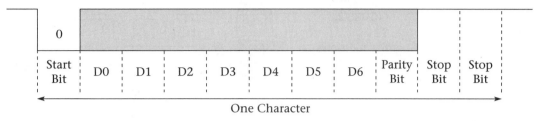

FIGURE 2.6 Asynchronous transmission

Synchronous Transmission

Some applications require transferring large blocks of data, such as reading a file from disk or transferring information from a computer to a printer. **Synchronous transmission** is an efficient method of transferring large blocks of data by using time intervals for synchronization.

One method of synchronizing the transmitter and the receiver is through the use of an external connection that carries a clock pulse. The clock pulse represents the data rate of the signal, as shown in Figure 2.7 and is used to determine the speed of data transmission. The receiver in Figure 2.6 reads the data as 01101.

FIGURE 2.7
Synchronous transmission

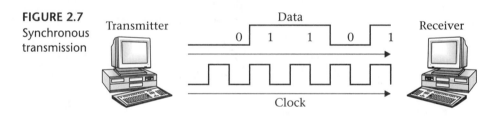

Figure 2.7 shows that an extra connection is required to carry the clock pulse for synchronous transmission. In networking one media is used for transmission of information and clock pulse. The two signals are encoded in such a way that the synchronization signal is embedded into the data. This can be done with Manchester encoding or Differential Manchester encoding.

2.6 Transmission Methods

There are two types of transmission methods used for sending digital signals from one station to another across a communication channel: serial transmission and parallel transmission.

Serial Transmission

In **serial transmission**, information is transmitted one bit at a time over one wire, as shown in Figure 2.8.

FIGURE 2.8
Serial
transmission

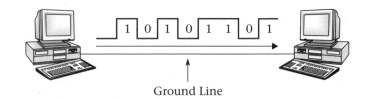

Ground Line

Parallel Transmission In **parallel transmission**, multiple bits are sent simultaneously, one byte or more at a time, instead of bit by bit as in serial transmission. Figure 2.9 shows how computer A sends eight bits of information to computer B at the same time by using eight different wires. Parallel transmission is faster than serial transmission, at the same clock speed.

FIGURE 2.9
Parallel
transmission

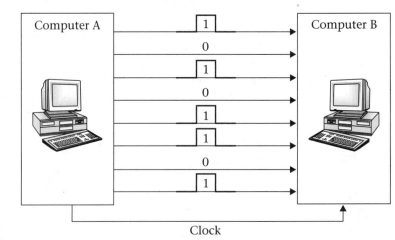

Clock

2.7 Communication Modes

A communication mode specifies the capability of a device to send and receive data by determining the direction of the signal between two connections. There are three types of communication modes: simplex, half-duplex and full-duplex.

Simplex Mode In **simplex mode**, transmission of data goes in one direction only, as shown in Figure 2.10. A common analogy is a commercial radio or TV broadcast—the sending device never requires a response from the receiving device.

FIGURE 2.10
Simplex
transmission

Sender   Receiver

Half-Duplex Mode In **half-duplex mode**, two devices exchange information as shown in Figure 2.11; however, information can be transmitted across the channel in only one direction at a time. A common example is Citizens' Band radio (CB) or ham radio—a user can either talk or listen, but both parties can not talk (or listen) at the same time.

FIGURE 2.11
Half-duplex
transmission

Transmitting in both directions,
but only one direction at a time

Full-Duplex Mode In **full-duplex mode**, both computers can send and receive information simultaneously, as shown in Figure 2.12. An example of full-duplex mode is our modern telephone system in which both users may talk and listen at the same time, with their voices carried two ways simultaneously over the phone lines.

FIGURE 2.12
Full-duplex
transmission

2.8 Bandwidth and Signal Transmission

Bandwidth is the measurement of the amount of information that can be transmitted in a fixed amount of time. For digital devices, bandwidth is expressed in bits per second or bytes per second. For analog devices, bandwidth is expressed in cycles per second.

Bandwidth The **bandwidth** of a communication signal is the range of frequencies that signal occupies. Therefore, the bandwidth is defined as:

$$\text{Bandwidth} = f_1 - f_2$$

where

f_2 is the lowest frequency of the signal and f_1 is the highest frequency of the signal

The lowest frequency of a human voice is 200 Hz and the highest frequency is 3300 Hz. Therefore, the bandwidth of the human voice = 3300 − 200 = 3100 Hz.

Signal Transmission There are two methods used to transfer information over media: baseband and broadband transmission.

Baseband Transmission Mode When the entire bandwidth of a cable is used to carry only one signal, the cable operates in **baseband mode**. Baseband transmission uses digital signals.

Broadband Transmission Mode When the bandwidth of a cable is used to carry several signals simultaneously, the cable operates in **broadband mode**. For example, cable television transmission works in broadband mode because it carries multiple channels using multiple signals over the same cable. Broadband mode uses analog signals.

2.9 Digital Signal Encoding

Digital signal encoding is used to represent binary values in the form of digital signals. The receiver of the digital signal must know the timing of each signal, such as the start and end of each bit. The following methods are used to represent digital signals:

- Unipolar Encoding
- Polar Encoding
- Bipolar Encoding
- Non-Return to Zero
- Non-Return to Zero Inverted (NRZ-I)
- Manchester and Differential Manchester Encoding

Manchester and Differential Manchester, and Non-Return to Zero Inverted encoding schemes are used in LANs and Non-Return to Zero is used in WANs. Each encoding technique is described below.

Unipolar Encoding In **unipolar encoding** only positive voltage or negative voltage are used to represent binary 0 and 1. For example, +5 volts represents binary 1 and 0 volts represents 0, as shown in Figure 2.13.

FIGURE 2.13
Unipolar encoding signals

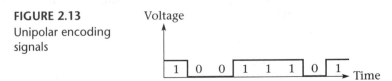

Polar Encoding In **polar encoding**, positive and negative voltages are used to represent binary 0 and 1. For example, +5 volts represents binary 1 and −5 volts represents binary 0, as shown in Figure 2.14.

FIGURE 2.14
Polar encoding
signals

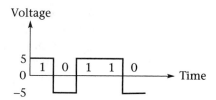

Bipolar Encoding   In bipolar encoding, signal voltage varies within three levels: positive, zero, and negative voltage. One of the most popular bipolar encoding methods is Alternate Mark Inversion (AMI).

In AMI encoding, binary 0 is represented by 0 volts and binary 1 alternates between positive and negative voltages, as shown in Figure 2.15.

FIGURE 2.15
Bipolar encoding
signals

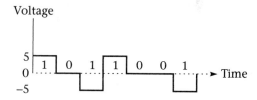

**Non-Return to
Zero Encoding
(NRZ)**   NRZ is the simplest signal encoding, using two voltage levels for representing 1 and 0, with binary 0 represented by positive voltage and binary 1 represented by negative voltage, as shown in Figure 2.16.

FIGURE 2.16
NRZ encoding
signals

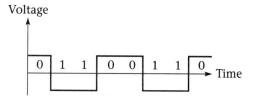

**Non-Return to
Zero Inverted
Encoding (NRZ-I)**   In NRZ-I there is a transition at the start of logic 1 (low to high or high to low) and no transition at start of 0, as shown in Figure 2.17.

FIGURE 2.17
NRZ-I encoding
signals

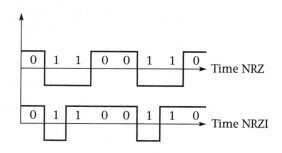

Manchester and Differential Manchester Encoding

In **Manchester** and **Differential Manchester encoding**, the clock pulse is embedded into the signal. Therefore, the receiver does not require any additional signal to represent the clock pulse. Table 2.5 shows how to convert digital signals to Manchester encoding and Differential Manchester encoding, and Figure 2.18 shows the parameters of Manchester and Differential Manchester encoding.

TABLE 2.5   Conversion Methods of Digital Signals to Manchester and Differential Manchester Encoding

Digital Signal	Manchester Encoding	Differential Manchester Encoding
Logic 1	Transition from high to low at the middle of original signal	Transition only in the middle of the signal
Logic 0	Transition from low to high at the middle of zero	Transition at the start of zero and also in the middle of zero (original signal)

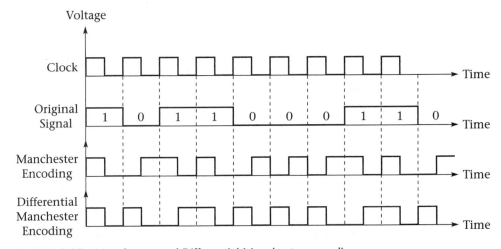

FIGURE 2.18   Manchester and Differential Manchester encoding

2.10   Error Detection Methods

When the transmitter sends a frame to the receiver, the frame can become corrupted due to external and internal noise. The receiver must first check the integrity of the frame before transmission begins. The possible sources of error are:

Impulse Noise A non-continuous pulse for a short duration is called **impulse noise**. It may be caused by a lightning discharge or a spike generated by the power switch being turned off and on.

Crosstalk This type of noise can be generated when a transmission line carrying a strong signal is coupled with a transmission line carrying a weak signal. The transmission line with the strong signal will produce noise (**crosstalk**) on the transmission line with the weak signal.

Attenuation When a signal travels on a transmission line, the strength of the signal is reduced over distance. This reduction is called **attenuation**. A weak signal is more affected by noise than a strong signal.

White Noise This type of noise exists in all electrical devices and is generated by moving electrons in the conductor. You may find it referred as either **white noise** or **thermal noise**.

The following methods can be used to detect an error:

- Parity Check or Vertical Redundancy Check (VRC)
- Block Check Character (BCC)
- One's Complement of the Sum
- Cyclic Redundancy Check (CRC)

Parity Check The simplest error detection method is the **parity check**. The parity check method can detect one error and is used in both the asynchronous transmission method and the character-oriented synchronous transmission method. A parity bit is an extra bit that the transmitter adds to the information before transmitting to the receiver. The value of parity bit selected by transmitter determines whether the data is given an even number of ones (even parity) or an odd number of ones (odd parity). For example: if a transmitter uses even parity to transmit the ASCII character 1000011 (lower case e), the transmitter adds parity bit 1 to the character in order that the number of the ones in the character becomes an even 11000011. The transmitter will then transmit 11000011 to the receiver. The receiver checks number of the ones in the character. If the number of the ones is even, there is no error in the character. Otherwise the character contains an error. Parity error detection is used in serial communication. Figure 2.19 shows the logic diagram for a parity bit generator using Exclusive-OR gates.

Block Check Character **Block Check Character (BCC)** uses vertical and horizontal parity bits in order to detect double errors. Remember, parity check is limited to detection of only one error and is used for transmitting single characters. When a block of characters is transmitted, BCC can be used to detect two errors. A parity bit is added to each character (row parity) in a block of characters, then column parity is computed. As illustrated in Table 2.6, the result of the column parity bits is called BCC. Table 2.6 shows that odd parity use for the rows and even parity use for columns results in a BCC of 0101110. For example, if two bits are changed in row one, such as B5 from 0 to 1 and B2 from 1 to 0, the row parity does not change; but the BCC will change, indicating the detection of two errors in one row.

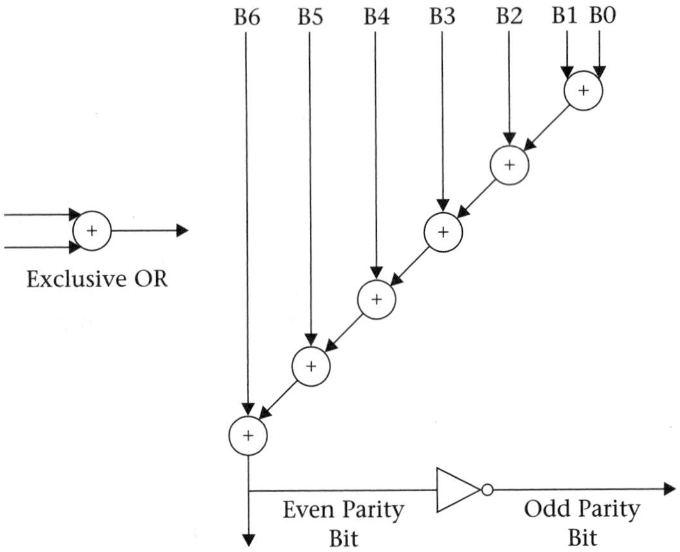

FIGURE 2.19
Logic diagram
of parity-bit
generator

Exclusive OR

Even Parity
Bit

Odd Parity
Bit

TABLE 2.6   BCC Calculations for "NETWORK"

Parity	B6	B5	B4	B3	B2	B1	B0	
1	1	0/1	0	1	1/0	1	0	N
0	1	0	0	0	1	0	1	E
0	1	0	1	0	1	0	0	T
0	1	0	1	0	1	1	1	W
0	1	0	0	1	1	1	1	O
0	1	0	1	0	0	1	0	R
1	1	0	0	1	0	1	1	K
0	1	0/1	1	1	1/0	1	0	BCC

One's Complement of the Sum The **One's Complement of the Sum** method is used for error detection of the Transmission Control Protocol (TCP) header and Internet Protocol (IP) header. At the transmitter side, the 16-bit one's complement of the sum of the header is calculated. The result of this calculation is transmitted with the information to the receiver. At the receiver side, the 16-bit one's complement of the header is calculated and compared to the result with the one's complement of the transmitter. If the two results are equal, there is no error. Otherwise there is error in the information. Figure 2.20 shows the one's complement of the sum for a four byte header.

FIGURE 2.20
One's complement
of the sum

Transmitting Side	Receiving Side
1000010	1000010
1111101 Header Contents	1111101
0000001	0000001
<u>0111100</u>	<u>0001100</u>
1111100	0011001
Therefore: One's Complement is 000011	Therefore: One's Complement is 1100110

Cyclic Redundancy Check

The parity bit and BCC can detect single and double errors. The **Cyclic Redundancy Check (CRC)** method is used for detection of a single error, more than a single error, and burst error (when two or more consecutive bits in a frame have changed).

The CRC uses modulo-2 addition to compute the Frame Check Sequence (FCS). In modulo-2 addition:

$1 + 1 = 0$, $1 + 0 = 1$, and $0 + 0 = 0$.

The following procedure is used to calculate FCS.
At the transmitter side:

Frame M is K bits.

P is a divisor of n + 1 bits.

FCS is n bits, and it is the remainder of $2^n * M/P$ using modulo-2 division.

At the transmitting side, the FCS, which is the remainder from dividing $2^n * M$ by P, is calculated. The transmitter will transmit frame $T = 2^n * M + FCS$ to the receiver, where T is K + n bits.

At the receiving side, the receiver divides T by P using modulo-2 division. If the result of this division generates a remainder of zero, there is no error in the frame. Otherwise the frame contains one or more errors.

Example 2.6

Find the Frame Check Sequence (FCS) for the following message. The divisor is given.

Message M = 111010, K = 6 bits
Divisor P = 1101, n + 1 = 4 bits

Therefore $2^n * M = 111010000$

By dividing $2^n * M$ by P using modulo-2 division, FCS = 010 as shown in Figure 2.21.

FIGURE 2.21
Frame Check
Sequence (FCS)
calculation

```
                101010 Quotient
        1101) 111010000
              1101
              001110
                1101
                001100
                  1101
                  00010
        Remainder or FCS is 010
```

At the transmitter side, the FCS is added to $2^n * M$, and the transmitter transmits frame T = 111010000 to the receiver.

At the receiver side, the receiver divides T by P, and if the result has a remainder of zero, there is no error in the frame. Otherwise the message contains an error. Since the above division takes time, special hardware is designed to generate FCS.

CRC Polynomial and Architecture   Below is a binary number represented by $b_5b_4b_3b_2b_1b_0$ where b_i represents each bit that can be represented by a polynomial:

$$b_5X^5 + b_4X^4 + b_3X^3 + b_2X^2 + b_1X + b_0$$

Example 2.7

Represent P = 1101101 by polynomial.

$$P(X) = X^6 + X^5 + X^3 + X^2 + 1$$

The following CRC polynomials are IEEE and ITU standards:

$$CRC\text{-}12 = X^{12} + X^{11} + X^3 + X^1 + 1$$
$$CRC\text{-}16 = X^{16} + X^{15} + X^2 + 1$$
$$CRC\text{-}ITU = X^{16} + X^{12} + X^5 + 1$$
$$CRC\text{-}32 = X^{32} + X^{26} + X^{23} + X^{22} + X^{16} + X^{11} + X^{10} + X^8 + X^7 + X^5 + X^4 + X^2 + X + 1$$

The CRC method uses a special Integrated Circuit (IC) to generate FCS. The design of this IC is based on the CRC polynomial. In general, a CRC polynomial can be represented by:

$$P(X) = X^n + \cdots + a_4X^4 + a_3X^3 + a_2X^2 + a_1X + 1$$

Figure 2.19 shows the general architecture of a CRC Integrated Circuit. C_i is a one bit shift register and the output of each register connected to input of Exclusive-OR gate; a_i is the coefficient of a CRC polynomial. In Figure 2.22, if a_i equals zero, then there is no connection between the feedback line and the XOR gate. In order to find FCS, the initial value for C_i is set to zero, and the message $2^n * M$ is shifted $K + n$ times through the CRC circuit. The final contents of C_{n-1}, ⋯ C_4, C_3, C_2, C_1, C_0 is the Frame Check Sequence.

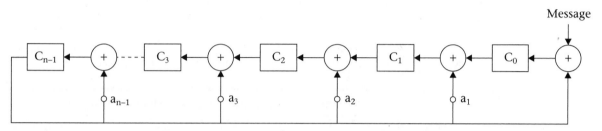

FIGURE 2.22   General Architecture of CRC Polynomial

Example 2.8

Show CRC circuit for polynomial:

$$P(X) = X^5 + X^4 + X^2 + 1$$

In the above polynomial the value for a_1, a_3 are zero and Figure 2.23 shows the CRC circuit for above polynomial

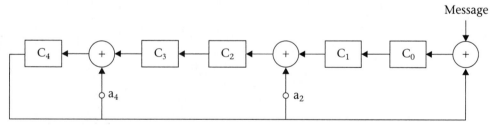

FIGURE 2.23   CRC circuit for polynomial $P(X) = X^5 + X^4 + X^2 + 1$

The following example shows how to find FCS by using CRC circuit:

For Message: $M = 111010$ and divisor $P = 1101$

The polynomial for $P = 1101$ is $P(X) = X^3 + X^2 + 1$. Using the general architecture for CRC, the circuit for $P(X)$ is shown in Figure 2.24, where $a_1 = 0$, $a_2 = 1$, and $a_3 = 1$.

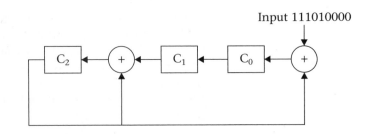

FIGURE 2.24
CRC circuit for
$P = 1101$

Input 111010000

Table 2.7 shows the contents of each register after shifting one bit at a time. After shifting 9 (K + n) times, the contents of the registers is FCS.

TABLE 2.7   FCS Value for Message: M = 111010 and P = 1101

C_2	C_1	C_0	Input
0	0	0	Initial Value
0	0	1	1
0	1	1	1
1	1	1	1
0	1	1	0
1	1	1	1
0	1	1	0
1	1	1	0
0	1	1	0
1	1	0	0
0	0	1	0
0	1	0	FCS

010 is Frame Check Sequence (FCS)

Summary

- Information transfer between two computers occurs in one of two types of signals: **digital** or **analog.**
- Modern computers work with digital signals.

- A Digital signal is represented by two voltages.
- **Binary** is the representation of a number in base-2.
- One digit in binary is called a **bit** and eight bits are equal to one **byte**. More than one byte is called a **word**.
- Information is represented inside the computer by binary or base-2.
- **Binary Coded Decimal (BCD)** is used for representing decimal numbers from 0 to 9.
- Information is represented by **ASCII Code** inside the computer; ASCII code is made up of seven bits.
- There are two methods used for transmission of data: **synchronous** and **asynchronous** transmission.
- **Parallel transmission** is a method by which data is transmitted byte by byte.
- **Serial transmission** is a method by which data is transmitted one bit at a time over a single transmission media.
- Asynchronous transmission adds extra bits (start bit and stop bit) to the character to achieve synchronization.
- Synchronous transmission uses a clock pulse for synchronization.
- There are three types of communication modes: **Simplex** transmission, **Half-Duplex** transmission, and **Full-Duplex** transmission.
- **Baseband** mode uses the bandwidth of a transmission media to carry only one signal.
- **Broadband** mode uses the bandwidth of a transmission media to carry multiple signals.
- Digital information can be represented by several forms of digital signal, such as **Non-Return to Zero (NRZ)**, **Non-Return to Zero Inverted (NRZ-I)**, **Manchester Encoding**, **Differential Manchester Encoding** and **Bipolar Encoding**.
- Some sources of error for digital communications are **impulse noise**, **crosstalk**, **attenuation**, and **white noise**.
- **Parity check**, **Block Check Character (BCC)**, **One's complement of sum** and **Cyclic Redundancy Checks (CRC)** are used for error detection in networking.

Key Terms

Analog Signal	Bandwidth
ASCII Code	Baseband Mode
Asynchronous Transmission	Binary
Attenuation	Bit

Block Check Character (BCC)
Broadband Mode
Byte
Crosstalk
Cyclic Redundancy Check (CRC)
Differential Manchester Encoding
Digital Signal
Full-Duplex Mode
Half-Duplex Mode
Impulse Noise
Manchester Encoding
Non-Return to Zero Encoding (NRZ)
Non-Return to Zero Inverted
   Encoding (NRZ-I)

One's Complement of the Sum
Parallel Transmission
Parity Check
Polar Encoding
Serial Tranmission
Simplex Mode
Synchronous Transmission
Thermal Noise
Unicode
Unipolar Encoding
White Noise
Word

Review Questions

● **Multiple Choice Questions**

1. Frequency (F) is the number of cycles in one second, and can be represented as _____.
   a. $F = 1/T$
   b. $F = T$
   c. $F = -1/T$
   d. $F = -T$

2. Modern computers work with _____ signals.
   a. digital
   b. analog
   c. a and b
   d. none of the above

3. Unicode is a new _____-bit character encoding standard code.
   a. 16
   b. 18
   c. 8
   d. 12

4. _____ transmission transmits data character by character.
   a. Asynchronous
   b. Synchronous
   c. Full-duplex
   d. Half-duplex

5. _____ transmission uses asynchronous transmission.
   a. Serial
   b. Parallel
   c. Broadband
   d. Full-duplex

6. In _____ mode, transmission of data goes in one direction only.
   - a. simple
   - b. half-duplex
   - c. full-duplex
   - d. serial

7. In _____ mode, both computers can send and receive information simultaneously.
   - a. simple
   - b. half-duplex
   - c. full-duplex
   - d. serial

8. The _____ of a communication signal is the range of frequencies that the signal occupies.
   - a. data rate
   - b. bandwidth
   - c. baud rate
   - d. broadband

9. What is the bandwidth of each computer for an Ethernet LAN with 20 computers? _____
   - a. 1 Mbps
   - b. 10 Mbps
   - c. 500 Kbps
   - d. 2 Mbps

10. Cyclic Redundancy Check can _____.
    - a. detect a single error and correct it
    - b. detect double errors and correct them
    - c. detect one or more than one error
    - d. correct one error or more than one error

11. Which of the following digital encodings is carry clock pulse? _____
    - a. Manchester Encoding
    - b. NRZ
    - c. RZ
    - d. RS-232

12. What is decimal value for $(111101)_2$? _____
    - a. 44
    - b. 63
    - c. 61
    - d. 52

13. What is hexadecimal value for $(111110111)_2$? _____
    - a. 1F6
    - b. 1F7
    - c. FB1
    - d. 7B6

14. The binary value for 45 is _____.
    - a. 101011
    - b. 101101
    - c. 101111
    - d. 011111

15. A range of frequencies carried by a medium is called _____.
    a. a broadband signal   c. an analog signal
    b. a baseband signal   d. a digital signal

16. Asynchronous communication uses _____.
    a. stop and start bits to indicate start of the character and end of the character
    b. a start bit to synchronize transmission
    c. start and stop bits for clocking
    d. none of the above

17. What is the efficiency of a serial connection using asynchronous transmission with 1 start bit, 2 stop bits and 7 data bits? _____
    a. 70%   c. 80%
    b. 75%   d. 65%

• Short Answer Questions
    1. Sketch an analog signal.
    2. What is frequency?
    3. What is the unit of frequency?
    4. What is the frequency of an analog signal that is repeated every .02 ms?
    5. Explain the amplitude of an analog signal.
    6. Sketch a digital signal.
    7. What is a bit?
    8. What is a byte?
    9. What is a word?
    10. Convert the following binary number to Hex.

        $(111000111001)_2 = ($ _____ $)_{16}$

    11. Convert the following binary numbers to decimal.

        $(11111111)_2 = ($ _____ $)_{10}$

        $(10110001)_2 = ($ _____ $)_{10}$

    12. Convert the following number to binary.

        $(FDE6)_{16} = ($ _____ $)_2$

    13. Convert the word DIGITAL to binary using the ASCII table.

14. Convert the word NETWORK to hexadecimal.

15. Write your name in binary ASCII, then change the result to hexadecimal.

16. What is serial transmission?

17. What is parallel transmission?

18. What is the advantage of parallel transmission over serial transmission.

19. Explain the following terms:
    a. Simplex
    b. Half-Duplex
    c. Full-Duplex

20. What is a synchronous transmission?

21. Why is a clock pulse needed for transmission of a digital signal?

22. Show the format of asynchronous transmission.

23. Sketch clock pulse.

24. List two types of digital encoding methods in which the clock is embedded in the data signal.

25. List methods of error detection.

26. List sources of error in networking.

27. Represent binary 110101 with polynomial.

28. Find the BCC for word "ETHERNET."

29. Show a CRC circuit for 1011.

30. Find the FCS for message 10110110 using the circuit in question 29.

31. Find the One's Complement of the Sum for the word "NETWORK."

32. Show the digital wave form for 0101011110.

33. Draw Manchester Encoding and Differential Manchester Encoding for binary 010110110.

34. Calculate the frequency of a signal repeated every 0.0005 second.

35. Calculate the bandwidth of a signal with the lowest frequency of 120 kHz and the highest frequency of 200 kHz.

36. Find the FCS for data unit 111011 with a divisor of 1011.

37. What is burst error?

38. What is the function of CRC polynomial?

chapter 3

Introduction to Computer Architecture

OBJECTIVES   After completing this chapter, you should be able to:

- List the components of a microcomputer and their functions
- List the components of a CPU
- Distinguish between a CPU and a microprocessor
- Compare a RISC processor and a CISC processor
- Discuss different types of memory
- Differentiate between various types of computer buses

INTRODUCTION   Just as the architecture of a building defines its overall design and function, so computer architecture defines the design and functionality of a computer system. The components of a microcomputer are designed to interact with one another, and this interaction plays an important role in the overall system operation.

3.1   Components of a Microcomputer

A standard microcomputer consists of a microprocessor (CPU), buses, memory, parallel input/output, serial input/output, direct memory access (DMA) and programmable I/O interrupt. Figure 3.1 shows the components of a microcomputer.

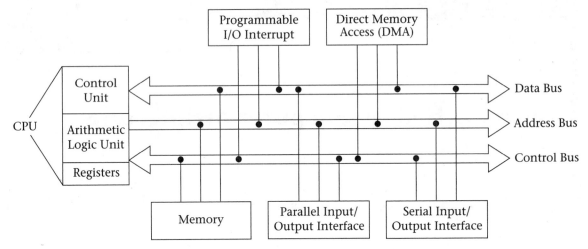

FIGURE 3.1 Components of a microcomputer

Central Processing Unit

The **central processing unit (CPU)** is the "brain" of the computer and is responsible for accepting data from input devices, processing the data into information, and transferring the information to memory and output devices. The CPU is organized into the following three major sections:

1. Arithmetic Logic Unit (ALU)
2. Control Unit
3. Registers

The function of the **arithmetic logic unit (ALU)** is to perform arithmetic operations such as addition, subtraction, division, and multiplication, and logic operations such as AND, OR, and NOT.

The function of the **control unit** is to control input/output devices, generate control signals to the other components of the computer such as read and write signals, and perform instruction execution. Information is moved from memory to the registers; the registers then pass the information to the ALU for logic and arithmetic operations.

It should be noted that the function of the microprocessor and CPU are the same. If the control unit, registers and the ALU are packaged into one integrated circuit (IC), then the unit is called a microprocessor, otherwise the unit is called a CPU. The difference in packaging is shown in Figure 3.2. There are two types of technology used to design a CPU: Reduced Instruction Set Computer (RISC) and Complex Instruction Set Computer (CISC).

CISC Architecture In 1978, Intel developed the 8086 microprocessor chip. The 8086 was designed to process a 16-bit data word; it had no instructions for floating-point operation. At the present time, the Pentium processes

FIGURE 3.2
Block diagram of
microprocessor
and CPU

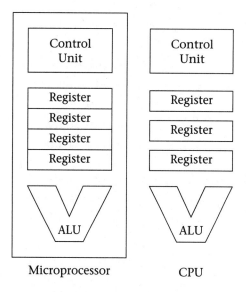

32-bit and 64-bit words and it can process floating-point instructions. Intel designed the Pentium processor in such a way that it can execute programs written for earlier 80x86 processors.

The characteristics of the 80x86 are called Complex Instruction Set Computers (CISC), which include instructions for earlier Intel processors. Another CISC processor is VAX 11/780, which can execute programs written for the PDP-11 computer. The CISC processor contains many instructions with different addressing modes, for example: VAX 11780 has more than 300 instructions with 16 different address modes.

The major characteristics of CISC processors are:

1. A large number of instructions
2. Many addressing modes
3. Variable length of instructions
4. Most instructions can manipulate operands in the memory
5. Control unit is microprogrammed

RISC Architecture   Until the mid-1990s, computer manufacturers were designing complex CPUs with large sets of instructions. At that time, a number of computer manufacturers decided to design CPUs capable of executing only a very limited set of instructions.

One advantage of reduced-instruction set computers is that they can execute their instructions very fast because the instructions are simple. In addition, the RISC chip requires fewer transistors than the CISC chip. Some of the RISC processors are the Power PC, MIPS RX000, IBM RISC System/6000, and SPARC.

The major characteristics of RISC processors are:

1. Require few instructions
2. All instructions are the same length (they can be easily decoded)
3. Most instructions are executed in one machine cycle
4. Control unit is hardwired
5. Few address modes
6. A large number of registers

Computer Bus When more than one wire carries the same type information, it is called a bus. The most common buses inside a microcomputer are the address bus, the data bus, and the control bus.

Address Bus The address bus defines the number of addressable locations in a memory IC by using the 2^n equation, where n represents the number of address lines. If the address bus is made up of three lines then there are $2^3 = 8$ addressable memory locations, as shown in Figure 3.3. The size of the address bus directly determines the maximum numbers of memory locations that can be accessed by the CPU.

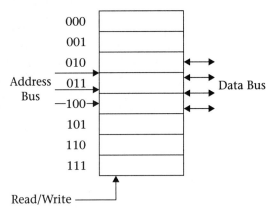

FIGURE 3.3
A RAM with three address lines and four data lines

Data Bus The data bus is used to carry data to and from the memory and represents the size of each location in memory. In Figure 3.3 each location can hold only four bits. If a memory IC has eight data lines, then each location can hold eight bits. The size of a memory IC is represented by $2^n \times m$ where n is the number of address lines and m is the size of each location. In Figure 3.3, where n = 3 and m = 4, the size of the RAM is:

$2^3 \times 4 = 32$ bits

Control Bus The control bus carries control signals from the control unit to the computer components in order to control the operation of each component. In addition, the control unit receives control signals from computer components. Some of the control signals are as follows:

Read Signal: The read line is used to read information from memory

Write Signal: The write line is used to write data into the memory

Interrupt: Indicates an interrupt request

Bus Request: The device is requesting to use the computer bus

Bus Grant: Gives permission to the requesting device to use the computer bus

I/O Read and Write: I/O read and write is used to read from or write to I/O devices

Memory In general, memory can hold information either temporarily or permanently. Following are some types of memory:

- Semiconductor Memory or Memory IC
- Floppy disk and Hard disk
- Tape
- CD-ROM (Compact Disk-Read Only Memory)

Semiconductor Memory There are two types of **semiconductor memory:** Random Access Memory (RAM) and Read only Memory (ROM).

Data can be read from or written into **Random Access Memory (RAM).** The RAM can hold the data as long as power is supplied to it. Figure 3.4 shows a general block diagram of RAM consisting of data bus, address bus, and read and write signals. The data bus carries information out of or into the RAM. The address bus is used to select a memory location. The read signal becomes active when reading data from RAM, and the write line becomes active when writing to the RAM. Remember, RAM can only hold information when it has power.

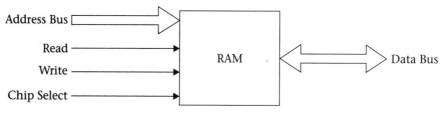

FIGURE 3.4 RAM block diagram

There are many types of RAM, such as **Dynamic RAM (DRAM),** Synchronous DRAM (SDRAM), and Static RAM (SRAM).

- DRAM is used in the main memory. It needs to be refreshed (recharged) about every 1 ms. The CPU cannot read from or write to memory while the DRAM is being refreshed—this makes DRAM the slowest running memory. A DRAM comes in different types of packaging such as the SIMM (Single In-Line Memory Module) and the DIMM (Dual In-Line Memory Module). The SIMM, shown in Figure 3.5, is a small circuit board that holds several chips. It has a 32-bit data bus. The DIMM is a circuit board that holds several memory chips. A DIMM has a 64-bit data bus.

FIGURE 3.5
DRAM SIMM

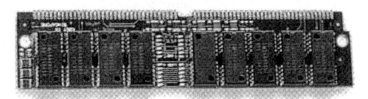

- **Synchronous DRAM (SDRAM)** technology uses DRAM and adds a special interface for synchronization. It can run at much higher clock speeds than DRAM. SDRAM uses two independent memory banks. While one bank is recharging, the CPU can read and write to the other bank. Figure 3.6 shows a block diagram of a SDRAM. Table 3.1 compares the throughput of SRAM and DRAM.

FIGURE 3.6
Block diagram
of SDRAM

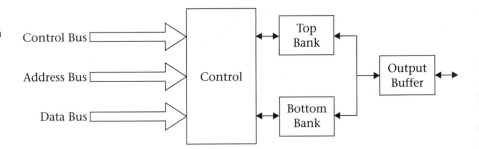

TABLE 3.1 Throughput of SDRAM Compared to DRAM

Bits Transferred	Two Banks SDRAM
1 bit	1.8 times DRAM
2 bit	2.4 times DRAM
8 bit	4.0 times DRAM
16 bit	4.4 times DRAM

- **Static RAM (SRAM)** is used in cache memory. SRAM is almost twenty times faster than DRAM and is also much more expensive.

Like its name suggests, information can be read only from the **Read Only Memory (ROM)**. ROM holds information permanently, even while there is no power to the ROM. Two types of ROM are listed below.

- **Erasable Programmable Read Only Memory (EPROM):** EPROM can be erased with ultraviolet light and reprogrammed with a device called an EPROM programmer.
- **Electrically Erasable PROM (EEPROM):** EEPROM can be erased by applying a specific voltage to one of the pins and can be reprogrammed with an EPROM programmer.

Parallel Input/Output Interface

The parallel I/O interface is used to connect parallel devices, such as printers and scanners, to the computer. Figure 3.7 shows Centronics parallel connectors.

FIGURE 3.7
Centronic male and female connectors

Centronics 36 pin male   Centronics 34 pin female

Serial Input/Output Interface

The serial I/O interface is used to connect serial I/O devices, such as a serial printer and modem, to the computer. The most common serial connector is an RS-232D. The RS-232 comes with a 25-pin connector, DB-25, and a 9-pin connector, DB-9, as shown in Figure 3.8. For pin assignment refer to Appendix A.

FIGURE 3.8
DB-25 and DB-9
serial connectors

Direct Memory Access

Direct Memory Access (DMA) allows for the transfer of blocks of data from memory to an I/O device or vice versa. Without DMA, the CPU reads data from memory and writes it to an I/O device. Transferring blocks of data from memory to an I/O device requires the CPU to perform one read and one write for each operation. This method of data transfer takes a lot of time. The function of DMA is to transfer data from memory to an I/O device directly, without using the CPU, so that the CPU is free to perform other functions.

The DMA performs the following functions in order to use the computer bus.

- The DMA sends a request signal to the CPU.
- The CPU responds to the DMA with a grant request, permitting the DMA to use the bus.

- The DMA controls the bus and the I/O device is able to read or write directly to or from memory.
- The DMA is able to load a file off a floppy disk into main memory when large blocks of data need to be transferred to a sequential range of memory. DMA is much faster and more efficient than a CPU.

Programmable I/O Interrupt

When multiple I/O devices such as floppy drives, hard disks, printers, monitors, and modems, are connected to a computer as shown in Figure 3.9, a mechanism is necessary to synchronize all of the device requests. The function of a programmable interrupt is to check the status of each device and inform the CPU of the status of each, for example: the printer is not ready, a disk is write protected, this is an unformatted disk, there is a missing connection to a modem. Each device sends a signal to the programmable I/O interrupt controller in order to update its status. Figure 3.9 shows the programmable I/O interrupt controller.

FIGURE 3.9
Programmable interrupt controller

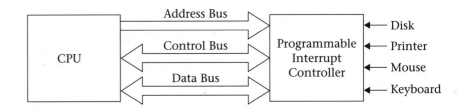

3.2 Memory Hierarchy

Computers come with three types of memory, which are arranged in a hierarchical fashion, as shown in Figure 3.10.

FIGURE 3.10
Memory hierarchy of microcomputer

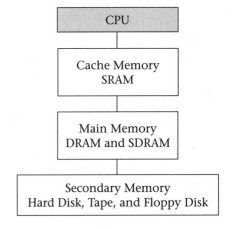

1. *Cache Memory:* **Cache memory** is the fastest type of memory and is most often used inside the microprocessor IC and on the motherboard. It is twenty times faster than main memory and therefore more expensive than main memory. Cache memory uses SRAM. Programs which reside in the main memory are divided into blocks of data, with some of the blocks moved to the cache; the CPU accesses the cache to get the data.

2. *Main Memory:* **Main memory** uses DRAM and SDRAM. The program for execution moves from secondary memory (disk or tape) into main memory.

3. *Secondary Memory:* **Secondary memory** refers to memory available on hard disk, tape, floppy disk, Zip drive, Jaz drive, and CD-ROM.

3.3 Disk Controller

The disk controller moves the head, reads and/or writes data, and determines the location of a file on the disk. Today there are two types of disk controllers: IDE (Integrated Disk Electronics) and SCSI (Small Computer System Interface).

Integrated Disk Electronics
An IDE disk drive is connected to the ISA bus with a flat ribbon cable. The IDE disk controller supports two hard disks, each with a 528-megabyte capacity. In 1994, hard disk drive vendors introduced EIDE (Extended IDE) which supports four devices, such as hard disks, tape drives, CD-ROM devices, and larger hard disk drives. The EIDE has two connectors. Each cable is connected to the EIDE controller and can support two hard disk drives with a capacity of up to 1 gigabyte. EIDE is used in IBM compatible computers.

Small Computer System Interface
The Small Computer System Interface (SCSI) standard is defined by the American National Standards Institute (ANSI) for connecting daisy chaining multiple I/O devices, such as scanner, hard disk, and CD-ROM to the microcomputer, as shown in Figure 3.11.

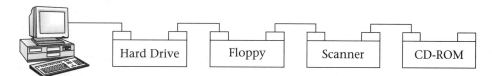

FIGURE 3.11 SCSI bus

SCSI is a standard interface for all types of microcomputers. It is used in Macintosh, RISC workstations, and minicomputers, as well as higher-end IBM compatible computers. The SCSI bus comes with different types of controllers. Table 3.2 shows the characteristics of several types of SCSI controllers, and Figure 3.12 shows SCSI-1, SCSI-2, and SCSI-3 connectors.

TABLE 3.2   Characteristics of Several Types of SCSI Controller

SCSI Connector	Bandwidth	Data Rate MB/s
SCSI-1	8 bits	5
SCSI-2	16 bits	10–20
Ultra SCSI	8 bits	20
SCSI-3	16 bits	40

MB/s Million Bytes per second

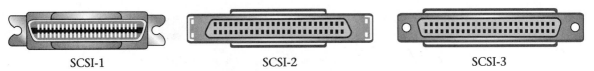

| SCSI-1 | SCSI-2 | SCSI-3 |

FIGURE 3.12   SCSI-1, SCSI-2 and SCSI-3 connectors

3.4   Microcomputer Bus

There are currently a number of different computer buses on the market that are designed for microcomputers. The following paragraphs provide a brief description of the ones most often used in microcomputers today.

ISA Bus   The **Industry Standard Architecture (ISA) bus** was introduced by IBM for the IBM PC using an 8088 Microprocessor. The ISA bus has an 16-bit data bus, and 20 address lines at a clock speed of 8 MHz. The PC AT type uses the 80286 processor which has a 16-bit data bus and 24-bit address lines and is compatible with the PC. Figure 3.13 shows the ISA card.

FIGURE 3.13
ISA card

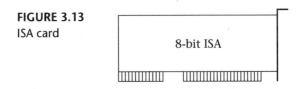

8-bit ISA

Microchannel Architecture Bus   The **Microchannel Architecture (MCA) bus** was introduced by IBM in 1987 for its PS/2 microcomputer. The MCA bus is a 32-bit bus that can transfer four bytes of data at a time and runs at 10-MHz clock speed. It also supports a 16-bit data transfer and has 32-bit address lines. Microchannel architecture was so expensive the non-IBM vendors developed a comparable but less expensive solution called the EISA bus.

EISA Bus   The **Extended ISA (EISA) bus** is a 32-bit bus that also supports 8- and 16-bit data transfer bus architectures. EISA runs at 8-MHz clock speeds and has 32-bit address lines. Figure 3.14 shows an EISA adapter card.

FIGURE 3.14
EISA card

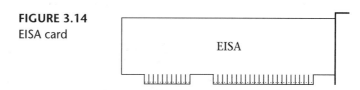

VESA Bus   The **Video Electronics Standard Association (VESA) bus,** which is also called a Video Local bus (VL-BUS), is a standard interface between the computer and its expansion. It provides faster data flow between devices.

PCI Bus   The **Peripheral Component Interconnect (PCI) bus** was developed by Intel Corporation. PCI bus technology includes a 32-bit bus that runs at 33-MHz clock speed. PCI offers many advantages for connections to hubs, routers, and Network Interface Cards (NIC). In particular, PCI provides more bandwidth, up to one gigabit per second as needed by these hardware components.

   The PCI bus was designed to improve the bandwidth and decrease latency in computer systems and was adopted by computer manufacturers. Current versions of the PCI bus support data rates of 1056 Mbps and can be upgraded to 4224 Mbps. The PCI bus can support up to 16 slots or devices in the motherboard. Most suppliers of ATM (Asynchronous Transfer Mode) and 100BaseT NICs offer PCI interface for their product. The PCI bus can be expanded to support a 64-bit data bus. Table 3.3 compares different bus architectures showing characteristics of ISA, EISA, MCA, VESA, and PCI buses. Figure 3.15 shows the PCI bus.

TABLE 3.3   Characteristics of Various Buses

Bus Type	ISA	EISA	MCA	VESA	PCI
Speed (MHz)	8	8.3	10	33	33/66
Data Bus Bandwidth (bits)	16	32	32	32	32/64
Max. Data Rate (MB/s)*	8	32	40	132	132/524
Plug and Play Capable	no	no	yes	yes	yes

* MB/s Megabytes/second

FIGURE 3.15
PCI card

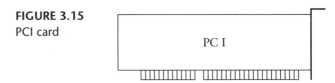

PC Card In the early 1990s, the rapid increase in the demand for mobile computing caused the development of smaller and more portable processing devices such as laptop computers. One of these developments was PC card technology. The almost spontaneous world-wide adaptation of the PC card was due in large part to the standard specification of the PC card by Personal Computer Memory Card International Association.

The PC card standard provides standards for three types of card: Type I, Type II, and Type III. Type I card is used for memory devices such as RAM, Flash Memory, and SRAM. Type II is used for I/O devices such as the fax and modem, and Type III is used for rotating mass storage devices.

The PC card adapter is covered by a metal casing and can be inserted and removed from a laptop computer at any time. This adapter is shown in Figure 3.16.

FIGURE 3.16
Inserting a
PC card

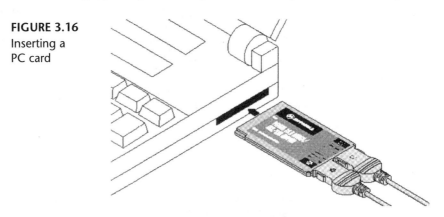

3.5   Plug-and-Play

When adding a new adapter to the computer, the Interrupt (IRQ), I/O address, and some DMA addresses must be set. Computers with an ISA bus must have these settings assigned manually by the user. The current EISA, MCA, and PCI bus technology can automatically assign the IRQ and I/O addresses. This automation process is called **plug-and-play** (PnP). Most new computers have the ability for plug-and-play, however, the adapter card itself must also have the capability to support plug-and-play. The adapter reports to the computer what IRQ and I/O addresses it is able to use, and the computer attempts to allocate the requested resources for the adapter. Operating systems such as Windows 98, Windows NT, and Macintosh OS support plug-and-play technology.

3.6 Universal Serial Bus

The **Universal Serial Bus (USB)** is a computer bus which enables users to connect peripherals such as the mouse, keyboard, modem, CD-ROM, scanner, and printer, to the outside of a computer without any configuration. Personal computers equipped with USB will allow the user to connect peripherals to the computer, and the computer will automatically be configured as the devices are attached to it. This means that a USB has the capability to detect when a device is added or removed from a PC. Up to 127 peripherals can be connected to a PC with a USB. The USB provides two data transfer rates: 1.5 Mbps and 12 Mbps.

Figure 3.17 shows the components of USB. The function of the serial interface engine (SIE) is to convert parallel data to serial data and vice versa. The function of the hub (repeater) is to accept information from the ports and transmit it to the SIE and conversely, to transmit data from the SIE to the ports. The maximum cable length for USB is five meters.

Figure 3.18 shows a USB cable which consists of four wires, with the V bus used to power the devices. The D+ and D– are used for signal transmission. In order to increase number of peripherals to the USB, several hubs are used as shown in Figure 3.19.

FIGURE 3.17
Architecture
of USB

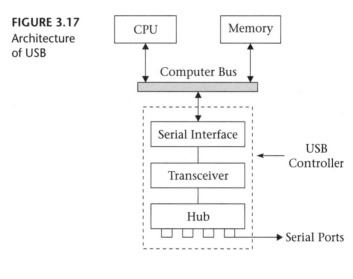

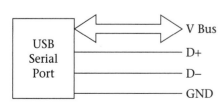

FIGURE 3.18 USB cable

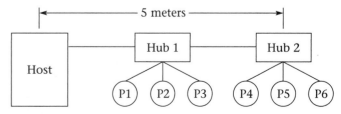

FIGURE 3.19 USB expansion

3.7 Intel Microprocessor Family

Intel designs and manufactures microprocessors for IBM compatible microcomputers. Each processor has a number or name, which is used by the computer designer to access the information provided by the manufacturer of the processor.

Intel microprocessor IC numbers and names are 8088, 80286 ,80386, 80486, Pentium, Pentium II and Pentium III. The Pentium III offers Single Instruction Multiple Data (SIMD) for floating point operations. The following is a list of the characteristics of a microprocessor:

1. *Register Size*: Registers are used to store information inside the processor. Register sizes can vary from 8- to 16- to 32- to 64-bit.

2. *Number of Registers*: A processor with several registers can store more information in the CPU for processing.

3. *Data Bus Size*: The size of the data bus is used for transferring data to or from memory or Input/Output ports.

4. *Address Bus Size*: The size of the address bus determines the number of memory locations that may be accessed by the microprocessor.

5. *Clock Speed*: The speed of the clock determines the speed at which a processor executes instructions.

6. *Math Coprocessor*: The math coprocessor is a special processor that performs complex mathematical operations.

7. *Real Mode*: Real mode allows for software compatibility with older software. It enables the processor to emulate the lowest Intel 8088 processor and use only the first 1 MB of memory.

8. *Protected Mode*: Protected mode is a type of memory utilization available on Intel 80286 and later model microprocessors. In protected mode, each program can be allocated a certain section of memory and other programs cannot use this memory. Protected mode also enable a single program to access more than 640K of memory.

9. *Cache Size*: Cache memory is a small amount of high speed memory used for temporary data storage based between the processor and main memory. The size of cache can help to speed up the execution time of a program.

10. *MMX Technology*: Intel's MMX technology is designed to speed up multimedia and communication applications, such as video, animation, and 3D graphics. The technology includes Single Instruction Multiple Data (SIMD) technique (meaning that with one instruction, the computer can perform multiple operations), 57 new instructions, eight 64-bit MMX registers and four new data types.

Table 3.4 provides a quick reference to the preceding list of characteristics.

TABLE 3.4 Characteristics of Intel Microprocessors

	80486dx	Pentium	Pentium Pro	Pentium Pro II	Pentium II
Register Size	32 bit	32 bit	32 bit	32/64 bit	32/64 bit
Data Bus Size	32 bit	64 bit	64 bit	64 bit	64 bits
Address Size	32 bit	32 bit	32 bit	32 bit	32 bits
Max. Memory	4 GB	4 GB	64 GB	64 GB	4 GB
Clock Speed	25, 33 MHz	60,166 MHz	150,200 MHz	233, 340, 400 MHz	450, 500 MHz
Math Processor	Built-in	Built-in	Built-in	Built-in	Built-in
L1 Cache	8 KB, 16 KB	8 KB Instruction 8 KB Data	8 KB Instruction 8 KB Data	16 KB Instruction 16 KB data	16 KB Instruction 16 KB data
L2 Cache	No	No	256 KB or 512 KB	512 KB	512 KB
MMX Technology	No	No	Yes	Yes	Yes

L1 Cache is the cache memory built inside the microprocessor.
L2 Cache is not part of the microprocessor; it is in a separate IC.

Summary

- The components of a computer are the **CPU, Memory, Parallel I/O, Serial I/O, Programmable Interrupt**, and **DMA**.
- The function of CPU is to process information using the Arithmetic Logic Unit.
- The components of a CPU are the Arithmetic Logic Unit (ALU), the Control Unit, and Registers.
- Most computers use three types of memory: **Cache Memory** (SRAM), **Main Memory** (DRAM or SDRAM) and **Secondary Memory** (hard disk, tape drive, and floppy disk).
- Semiconductor memory types are: **SRAM, DRAM, SDRAM, ROM**, and **EPROM**.
- SCSI-1, SCSI-2 and SCSI-3 are computer peripheral controllers.
- **ISA, EISA, MCA**, and **PCI** are microcomputer buses.
- The **Universal Serial Bus** is used to connect peripherals to the PC so that PC can automatically detect a device.

Key Terms

Arithmetic Logic Unit (ALU)
Cache Memory
Central Processing Unit (CPU)

Complex Instruction Set Computer (CISC)
Control Unit

Direct Memory Access (DMA)

Dynamic RAM (DRAM)

Electrically Erasable PROM (EEPROM)

Erasable Programmable Read Only
Memory (EPROM)

Extended ISA (EISA) Bus

Industry Standard Architecture (ISA) Bus

Intel Microprocessor Family

MMX Technology

Main Memory

Microchannel Architecture (MCA) Bus

Peripheral Component Interconnect
(PCI) Bus

Plug-and-Play

Random Access Memory (RAM)

Read Only Memory (ROM)

Reduced Instruction Set Computer
(RISC)

Secondary Memory

Single In-Line Memory Module (SIMM)

Small Computer System Interface (SCSI)

Static RAM (SRAM)

Synchronous Dram (SDRAM)

Universal Serial Bus (USB)

Video Electronics Standards
Association (VESA) Bus

Review Questions

● **Multiple Choice Questions**

1. The function of the _____ is to perform arithmetic operations.
   a. bus
   b. serial port
   c. ALU
   d. control unit

2. When you compare the functions of a CPU and a microprocessor, _____.
   a. they are the same
   b. they are not the same
   c. the CPU is faster than the microprocessor
   d. the microprocessor is faster than the CPU

3. RISC processors are _____.
   a. complex instruction set
   b. reduced instruction set
   c. a and b
   d. none of the above

4. CISC processor control unit is _____.
   a. hardwired
   b. microcode
   c. a and b
   d. none of the above

5. Which of the following memory types are used for main memory? _____
   a. ROM and SDRAM
   b. SRAM and DRAM
   c. SDRAM and DRAM
   d. DRAM and ROM

6. _____ holds information permanently, even while there is no power.
   a. ROM   c. RAM
   b. DRAM   d. SRAM

7. Direct memory access allows for the transfer of blocks of data from memory to an I/O device (or vice versa) without using the _____.
   a. CPU   c. control bus
   b. data bus   d. DMA controller

8. _____ is the fastest type of memory.
   a. Cache memory   c. Secondary memory
   b. Main memory   d. Hard disk

9. Which of the following buses are 32-bit? _____
   a. ISA and PCI
   b. PCI and EISA
   c. EISA and MCA
   d. MCA and ISA

10. Which of the following operating systems supports plug-and-play? _____
    a. Windows NT and Windows 95
    b. Windows 98 and Windows NT
    c. Windows 95 and Windows 98
    d. DOS and Windows NT

● **Short Answer Questions**
   1. List the components of a microcomputer.
   2. Explain the functions of a CPU.
   3. List the functions of an ALU.
   4. What is the function of a control unit?
   5. Distinguish between a CPU and a microprocessor.
   6. What does RAM stand for?

7. What is SRAM and discuss its application?

8. Define DRAM and SDRAM and explain their applications.

9. Explain the function of an address bus and a data bus.

10. What does IC stand for?

11. What is the capacity of a memory IC with 10 address lines and 8 data lines?

12. What is ROM?

13. What does EEPROM stand for, and what is its application?

14. What is SIMM?

15. Explain the function of cache memory and give its location.

16. List the types of memory used for main memory.

17. List the types of memory used for secondary memory.

18. Explain the function of DMA.

19. What is the application of a parallel port?

20. What is the application of a serial port?

21. What is an interrupt?

22. Explain plug-and-play.

23. List the operating systems which support plug-and-play.

24. What does SCSI stands for?

25. What does IDE stands for?

26. List some of the computer buses.

27. Explain the difference between CISC processors and RISC processors.

chapter 4

Standards Organizations and OSI Model

OBJECTIVES
After completing this chapter, you should be able to:

- List some of the organizations responsible for developing standards for networks and data communication
- Discuss the concepts of communication protocols
- Explain the OSI model for networking and the function of each layer
- Comprehend frame transmission methods
- Demonstrate your understanding of error and flow control
- List some of the IEEE 802 Committee standards
- Draw the Logical Link Control (LLC) frame format

INTRODUCTION
There are several organizations that are constantly working toward developing standards for computers and other communication equipment. The development of standards for computers could eventually enable hardware and software products made by different vendors to be compatible. Standardization would allow products from different manufacturers to work together in creating customized systems. Without standards, only hardware and software made by the same manufacturer can be combined with certainty that it will work properly. The following is a list of **standards organizations**:

    IEEE: The Institute for Electrical and Electronics Engineers (IEEE) is the largest technical organization in the world. The objective of IEEE is to advance the field of electronics, computer science, and computer engineering. IEEE also develops standard for computers, electronics, and local area networks (in particular, the IEEE 802 standards for the LAN).

ITU: The International Telecommunication Union (ITU) was founded in 1864 and became a United Nations Agency with the purpose of defining standards for telecommunications, Wide Area Networks (WAN), Asynchronous Transfer Mode (ATM), and Integrated Services Digital Networks (ISDN).

EIA: The Electrical Industry Association (EIA) is a trade association representing high technology manufacturers in the United States. The EIA develops standards for connectors and transmission media. Some of the well-known EIA standards are RS-232 and RJ-45.

ANSI: The American National Standards Institute (ANSI) was founded in 1918. ANSI is composed of 1300 members representing computer companies, with the purpose of developing standards for the computer industry.

ANSI is the United States representative in the International Organization for Standardization (ISO). Some of the well-known ANSI standards are optical cable, programming language (ANSI C), and Fiber Distributed Data Interface (FDDI).

ISO: The International Organization for Standardization. (ISO) is an international organization comprising the national standards of seventy-five countries. ISO develops standards for a wide range of products, including the model for networks called the Open System Interconnection (OSI) model.

IETF: The Internet Engineering Task Force (IETF) develops standards for the internet, such as Internet Protocol version 6 (IPv6). IETF is composed of international network designers, network industries, and researchers.

4.1 Communication Protocols

A **communication protocol** is a set of rules used by computers, which allows them to communicate with each other. Computers must follow certain rules in order to able to communicate with each other. Some of the rules a protocol must define are:

- *Size of information.* Both computers must agree on the minimum and maximum size of information.
- *How to represent information.* Information may be Unicode, ASCII, or encrypted.
- *Error detection.* The method used by the receiver to check the integrity of information must be established.
- *Receipt of information.* The transmitter must know that information has been received at the destination.
- *Non-receipt of information.* Both computers must know what to do if information sent is not received or if it is received but has been corrupted.

Some of the common network protocols are:

TCP/IP: Transmission Control Protocol/Internet Protocol; used in the Internet.

NetBEUI: NetBIOS Extended User Interface is a small and fast protocol used for small LAN.

X.25: X.25 is a set of protocols used in packet switching networks.

IPX/SPX: Novell NetWare uses Internet Packet Exchange/Sequenced Packet Exchange.

NWlink: NWlink is a Microsoft version of IPX/SPX.

4.2 Open System Interconnection Model

The **Open System Interconnection (OSI) model** was developed by the International Organization for Standardization for interoperability between equipment designed for networks. The ISO developed this open system reference model for networking. Any device which meets the OSI standards can be easily connected to any other device that adheres to the OSI model. An open system is a set of protocols that allows two computers to communicate with each other regardless of their design, manufacturer, or CPU type. The OSI model divides network communication into seven layers, with each layer performing specific tasks, as shown in Figure 4.1.

Layer 1: Physical Layer

The **Physical layer** defines the type of electrical signal, and type of connectors (such as RS-232 or RJ-45) to be used for the Network Interface Card (NIC). It defines cable types (such as coaxial cable, twisted-pair, or fiber-optic cable) to be used for transmission media.

Layer 2: Data Link Layer

The **Data Link layer** defines the frame format, such as start of frame, end of frame, size of frame, and type of transmission. The Data Link layer performs the following functions:

1. *On the transmitting side:* The Data Link layer accepts information from the Network layer and breaks the information into frames. It then adds the destination address, source address, Frame Check Sequence (FCS) field, and length field to each frame and passes each frame to the Physical layer for transmission.

2. *On the receiving side:* The Data Link layer accepts the bits from the Physical layer and forms them into a frame, performing error detection. If the frame is free of error, the Data Link layer passes the frame up to the Network layer.

3. *Frame synchronization:* It identifies the beginning and end of each frame.

4. *Flow control*

5. *Distinguishes between control frame and information frame*

6. *Link management:* It coordinates transmission between the transmitter and the receiver.

Layer 7	**Application Layer** Performs information processing such as file transfer, e-mail, and Telnet.

Layer 6	**Presentation Layer** Defines the format of data to be sent: ASCII, data encryption, data compression, and EBCDIC.

Layer 5	**Session Layer** Sets up a session between two applications by determining the type of communication such as duplex, half-duplex, synchronization, etc.

Layer 4	**Transport Layer** Ensures data gets to the destination. Manages error control, flow control, and quality of the service.

Layer 3	**Network Layer** Sets up connection, disconnects connection, provides routing and multiplexing.

Layer 2	**Data Link Layer** Manages framing, error detection, and retransmission of message.

Layer 1	**Physical Layer** Electrical Interface (type of signal), Mechanical interface (type of connector). Converts electrical signals to bits, transmits and receives electrical signals.

FIGURE 4.1 OSI model

There are several existing protocols for the Data Link layer such as:

- **Synchronous Data Link Control (SDLC):** SDLC was developed by IBM as link access for System Network Architecture (SNA).
- **High Level Data Link Control (HDLC):** HDLC is a version of SDLC modified by the ISO for use in the OSI model.
- **Link Access Procedure Balanced (LAPB):** HDLC was modified by ITU, and it is called LAPB.

High Level Data Link Control   High Level Data Link Control (HDLC) is designed to work with any type of station, such as primary, secondary, or combined station (a combination of primary and secondary stations). The function of a **primary station** is to control the network and thus be able to transmit information at any time.

A **secondary station** can transmit information only when requested by a primary station, Figure 4.2(a) shows a point-to-point link using primary and secondary stations and Figure 4.2(b) shows multipoint links between primary and secondary stations.

FIGURE 4.2(a)
Point-to-point connection between primary and secondary stations

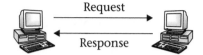

FIGURE 4.2(b)
Multipoint connection using one primary station and several secondary stations

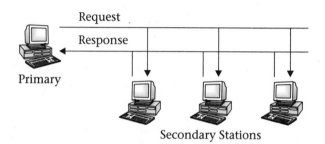

A **combined station** operates as both a primary and a secondary station in one unit. It can request data from other stations and can also respond to the requests of other stations. Figure 4.2(c) shows a point-to-point link between two combined stations.

FIGURE 4.2(c)
Point-to-point connection for combined stations

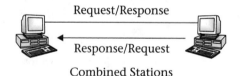

HDLC Frame Format   Figure 4.3 shows the general frame format of the HDLC, where the flag field is 01010101. The address field gives the address of destination and the control field determines the type of information in the information field such as Information Frame (I-Frame), Supervisory Frame (S-Frame), and Unnumbered Frame (U-Frame).

FIGURE 4.3
HDLC frame format

1	1 or more	1 or 2	variable	2 or 4	1 byte
Flag	Address	Control	Information	FCS	Flag

Information Frame (I-Frame)   Figure 4.4(a) shows the control field for the information frame (I-Frame)

FIGURE 4.4(a)
Control field for
I-frame

where

N(S) is the sequence number of a transmitted frame,

N(R) is the sequence number of the next frame expected to receive,

P/F (Poll/Final) = 0 implies this frame is the last frame, and

P/F = 1 indicates the primary station is requesting a secondary station to send its frame.

Supervisory Frame (S-Frame)   An S-Frame is used for flow and error control such as receiver ready, receiver not ready, and receiver reject. The code field in Figure 4.4(b) defines the supervisory frame.

FIGURE 4.4(b)
Control field for
S-frame

Unnumbered Frame (U-Frame)   Figure 4.4(c) shows the control field for a U-Frame. The U-Frame is used for setting the mode of operation between source and destination, disconnecting a logical link, resetting a connection, and testing a connection. The code field distinguishes the U-Frame from the S-Frame.

FIGURE 4.4(c)
Control field for
U-frame

Layer 3: Network Layer   The function of the **Network layer** is to perform routing. Routing determines the route or pathway for moving information (in a network with multiple LANs). The Network layer checks the logical address of each frame and forwards the frame to the next router based on a lookup table. The Network layer is responsible for translating each logical address (name address) to a physical address (MAC address).

The Network layer provides two types of services: connectionless and connection-oriented services. In connection-oriented service, the Network layer makes a connection between source and destination, then transmission starts. In connectionless service, there is no connection between source and destination. The source transmits information regardless of whether the destination is ready or not. A common example of this type of service is e-mail.

Layer 4:
Transport
Layer

The **Transport layer** provides for the reliable transmission of data in order to ensure that each frame reaches its destination. If, after a certain period of time, the Transport layer does not receive an acknowledgment from the destination, it retransmits the frame and again waits for acknowledgment from the destination.

Layer 5: Session
Layer

The **Session layer** establishes a logical connection between the applications of two computers that are communicating with each other. It allows two applications on two different computers to establish and terminate a session. When a workstation connects to a server, the server performs the login process, requesting a username and password. This is an example of establishing a session.

Layer 6:
Presentation
Layer

The **Presentation layer** receives information from the Application layer and converts it to a form acceptable by the destination. The Presentation layer converts information to ASCII or Unicode, or encrypts or decrypts (decryption) the information.

Layer 7:
Application
Layer

The **Application layer** enables users to access the network with applications such as e-mail, FTP, and Telnet.

4.3　Frame Transmission Methods

The Data Link layer offers two types of frame transmission: asynchronous transmission and synchronous transmission.

Asynchronous
Transmission

In **asynchronous transmission**, each character is transmitted separately, as shown in Figure 4.5. One disadvantage of this method is that extra bits must be added to each group of character bits, such as start bits and stop bits, which represent the beginning and end of each character, and parity bits, which are used for error detection. This method is inefficient for transferring a large volume of information. The modem is the primary application that uses asynchronous protocol.

Start Bit	Character Bits	Parity Bit	Stop Bits		Start Bit	Character Bits	Parity Bit	Stop Bits

FIGURE 4.5　Asynchronous transmission

Synchronous
Transmission

Synchronous transmission is more efficient than asynchronous transmission for transmitting a large block of data. Synchronous transmission can be oriented to either characters or bits.

Character Synchronization   **Character-oriented synchronization** is used for character-oriented transmissions in which blocks of data, such as text files in ASCII format, are transferred over a network. Figure 4.6 shows the frame format for character-oriented transmission.

FIGURE 4.6
Character
synchronization

◀────── Direction of Information Flow

SYN	SYN	STX	Information	ETX

The following are functions of each field in the Character synchronization frame format:

SYN: The Synchronization Character is used by the receiving device to synchronize its timing. The SYN character is 16 in Hex, with two SYNs in each frame.

STX: The Start of Text (STX) field is used by the receiver in order to recognize the start of data. STX is 02 in Hex.

ETX: The End of Text (ETX) field is used to inform the receiver that the end of the data has been reached. ETX is 03 in Hex.

Data in the information field is in the form of printable characters. Since the information field might contain STX or ETX characters, the transmitter precedes the STX or ETX with a DLE character as shown in Figure 4.7. Therefore the receiver will interpret a DLE STX as the start of frame and a DLE ETX as the end of the frame. If the information field contains a DLE character, the transmitter sends DLE DLE. When it sees a DLE DLE sequence, the receiver knows this DLE does not belong to any of start-of-text or end-of-text sequence. The receiver then strips off one of the DLE characters within the information block.

FIGURE 4.7
Character synchronization
with DLE character

SYN	SYN	DLE STX	DLE DLE	DLE ETX

Bit-Oriented Synchronization   Since the character-oriented synchronization technique used in synchronous protocol needs several control characters, it is less efficient than bit synchronization.

Bit-oriented synchronization is used for bit-oriented transmission—when information is transmitted bit by bit. Figure 4.8 shows the frame format of a bit synchronization frame. The start of the frame and the end of the frame are represented by eight bits in the form of 01111110. If the information field happens to contain the binary value 01111110, which normally indicates the end of a frame, an adjustment must be made by the transmitter. The transmitter inserts an extra zero

FIGURE 4.8
Format of bit-oriented synchronization

0 1 1 1 1 11 0	1101100000111111	0 1 1 1 1 10 10
Start Flag	Information Field	End Flag

after five ones are repeated in the information field. The receiver will discard this extra zero. This technique is called bit insertion, as shown in Figure 4.9.

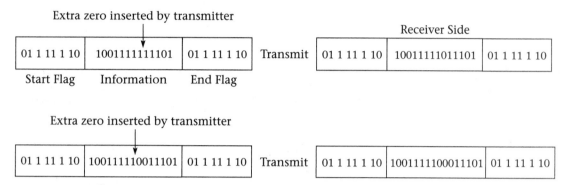

Extra zero inserted by transmitter

01 1 11 1 10	1001111111101	01 1 11 1 10
Start Flag	Information	End Flag

Receiver Side

Transmit | 01 1 11 1 10 | 10011111011101 | 01 1 11 1 10 |

Extra zero inserted by transmitter

01 1 111 1 10	100111110011101	01 1 111 1 10

Transmit | 01 1 111 1 10 | 1001111100011101 | 01 1 111 1 10 |

FIGURE 4.9 Bit insertion in information field

4.4 Error and Flow Control

Functions of Data Link layer are error detection, error control, and flow control. During the transmission of a frame from a source to its destination, the frame may get corrupted, or lost. It is the function of the Data Link layer of the destination to check for error in the frame and inform the source about the status of the frame. This function must be performed in order for the source to retransmit the frame. One of the methods used is **Automatic Repeat Request** (ARQ): positive or negative acknowledgment is used to establish reliable communication between the source and the destination. Automatic repeat request is carried out in two ways (stop-and-wait ARQ and Go-Back-N ARQ).

Stop-and-Wait ARQ In Stop-and-Wait ARQ the source transmits a frame and waits for a specific time for acknowledgment from the destination. If the source does not receive acknowledgment during this time, the source retransmits the frame. This method is used for a network with half-duplex connection.

Case 1: The source station transmits a frame to the destination station. The destination station checks the frame for any errors. If there is no error in the frame, the destination station responds to the source station with a Positive Acknowledgment Frame ACK(N), where N is the sequence number of the frame. The source station then transmits the next frame, as shown in Figure 4.10.

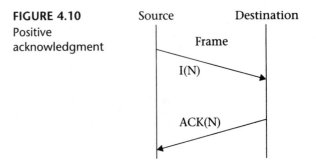

FIGURE 4.10
Positive
acknowledgment

Case 2: The source station transmits a frame to the destination station. The destination station checks the frame for error. If there is an error in the frame, the destination station responds to the source station with a Negative Acknowledgment Frame (NACK). Then the source retransmits the frame, as shown in Figure 4.11.

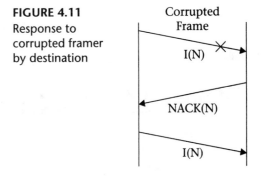

FIGURE 4.11
Response to
corrupted framer
by destination

Case 3: The source station transmits a frame to the destination station. The source did not receive any acknowledgment due to the loss of the frame or loss of acknowledgment from the destination. When the source starts to transmit a frame to the destination, it sets a timer and waits for acknowledgment. If the source does not receive an acknowledgment from destination during that period of time, the frame is retransmitted, as shown in Figure 4.12.

Continuous ARQ   In continuous ARQ the transmitter continues to transmit frames to the destination. The destination sends ACK or NACK on different channels. The continuous ARQ is used in a packet-switching network with the full-duplex connection. There are two types of continuous ARQ.

Go-Back-N ARQ   In the Go-Back-N ARQ method the transmitter continues to transmit and the receiver to acknowledge for each frame in a different channel, as shown in Figure 4.13.

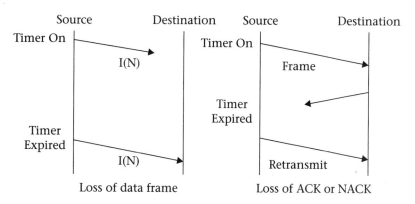

FIGURE 4.12   Loss of ACK or NACK and I-frame

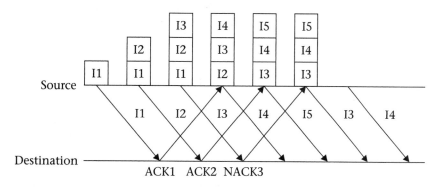

FIGURE 4.13   Go-back-N ARQ

Figure 4.13 shows that the source transmitted frame I5 and received NACK from I3; the source will retransmit frames I3, I4, and I5. In Go-Back-N, the source should hold a copy of those frames not receiving acknowledgment. When the source receives acknowledgment for a frame, it can remove the frame from its buffer.

Selective Reject ARQ   In selective ARQ the source will retransmit only those frames for which the destination has sent a negative acknowledgment. Figure 4.14 shows the source transmitted frame I3 and received NACK1, which indicated frame I1 was corrupted. The source retransmits only frames I1. In this method, the destination should have the capability to reorder frames which are out of order.

Sliding Window Method   In continuous ARQ the source keeps a copy of the transmitted frames in its buffer until it receives acknowledgment for a frame; it then removes the frame from its buffer. The continuous ARQ has the following deficiencies:

FIGURE 4.14
Selective Reject ARQ

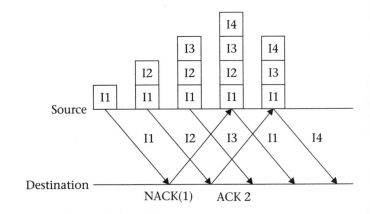

1. The destination may not have enough memory to store incoming frames.
2. The source may transmit frames faster than the destination can process them.
3. The source must hold a copy of all unacknowledged transmitted frames in its buffer; therefore, the source requires a large buffer.
4. The file to be transmitted is divided into packets. Each packet has a sequence number; if the sequence number becomes large, it decreases network efficiency.

The Sliding Window Method limits the number of frames waiting for acknowledgment in the source. For example, a source with a window of seven means the source can hold only seven unacknowledged frames in its buffer. The source will stop transmitting after seven unacknowledged frames and wait for an acknowledgment frame. When the source receives acknowledgment for a frame, it removes the frame from its buffer and transmits the next frame.

In order to avoid using a large sequence number, most of the networking protocol uses the following formula for assigning a sequence number to each frame.

Sequence Number = Frame Number Modulo K

where

K is the window size of the sending station

For example: What is the sequence number for frame number 25? Assume the window size is 7.

Sequence Number = 25 Modulo 7 = 4

4.5 IEEE 802 Standard Committee

The **IEEE 802 Committee** defined the standard for the Physical layer and the Data Link layer in February of 1980 and named it IEEE 802, with "80" representing 1980 and "2" representing the month of February. Figure 4.15 shows the IEEE 802 standard and the OSI model. The IEEE standard divides the Data Link layer of OSI Model to two sublayers: Logical Link Control (LLC) and Media Access Control (MAC).

FIGURE 4.15
IEEE standard model and OSI model

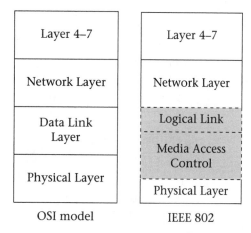

OSI model   IEEE 802

Media Access Control The **Media Access Control (MAC)** layer defines the method that stations use to access the network, such as:

- Carrier Sense Multiple Access with Collision Detection (CSMA/CD)—used for Ethernet
- Control Token—used in Token Ring Network and Token Bus Network

Logical Link Control The **Logical Link Control (LLC)** defines the format of the frame. It is independent of network topology, transmission media, and Media Access Control.

Figure 4.16 shows different MAC layers for several IEEE 802 networks. All networks which are listed use the same logical link control. Figure 4.17 shows the frame format of the LLC, which is used by all IEEE 802.X projects.

The following are the functions of each field of the LLC frame format:

Destination Service Access Point (DSAP): Since the destination station might run several network protocols such as Novell Netware, NetBIOS, Windows NT, and TCP/IP, the DSAP has to show the address of the protocol for the destination. Table 4.1 shows the most common value for service access point (SSAP and DSAP).

Source Service Access Point (SSAP): SSAP is a value of the source protocol and indicates the protocol that was used by the transmitter to send the packet.

Control Field: The control field determines what type of information is stored in the information field, such as information frame, supervisory

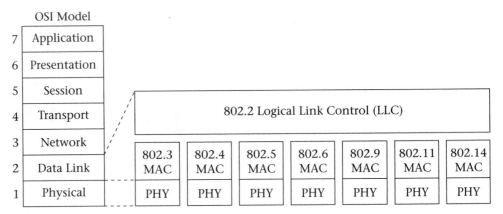

FIGURE 4.16 IEEE 802 reference model

FIGURE 4.17
Logical link control
frame format

1 byte	1 byte	1 or 2 bytes	
DSAP	SSAP	Control	Information

TABLE 4.1 DSAP and SSAP Values

Protocol	SSAP and DSAP Values
IBM SNA	04
IP	06
3Com	80
Novell	E0
Banyan	BC
NetBIOS	F8
LAN Manager	F4

frame, and unnumbered frame. The supervisory frames are: receiver ready, receiver not ready, and reject. Some of the unnumbered frames are: reset, frame reject, disconnect, and set asynchronous respond mode.

Following is a partial list of standards developed for networking by IEEE 802 Committee:

802.1   Internetworking
802.2   Logical Link Control (LCC)

802.3	Ethernet
802.4	Token Bus LAN
802.5	Token Ring LAN
802.6	MAN
802.7	Broadband Technical Advisory Group
802.8	Fiber-Optics Technical Advisory Group
802.9	Integrated Services Data Networks (ISDN)
802.10	Network Security
802.11	Wireless Networks
802.12	100 VG–AnyLAN
802.14	Cable Modem

Summary

- Some of the **standards organizations** which develop standards for networks and data communications are: The IEEE, ITU, EIA ISO, IETF, and ANSI.

- A **communication protocol** is a set of rules used by two computers in order to communicate which each other.

- The most popular communication protocols are Transmission Control Protocol/ Internet Protocol (TCP/IP), NetBEUI, IPX/SPX, NWlink, and DecNet.

- The **International Organization for Standardization (ISO)** developed a model for networks called the **Open System Interconnection (OSI) model**.

- The OSI model consists of seven layers, listed from top to bottom: Application Layer, Presentation Layer, Session Layer, Transport Layer, Network Layer, Data Link Layer and Physical Layer.

- The **Application Layer** enables the user to access the network.

- The **Presentation Layer** is responsible for representation of information such as ASCII, encryption, and decryption.

- The function of **Session Layer** is to establish a session between the source application and destination application, and to disconnect a session between two applications.

- The function of **Transport Layer** is to ensure that data gets to the destination, to perform error control and flow control, and to assure quality of service.

- The function of **Network Layer** is to deliver information from the source to the destination and route the information.

- The function of **Data Link Layer** is framing, error detection, and retransmission.

- The functions of **Physical Layer** are: electrical interface, type of the signal, and conversion of bits to signals (electrical or optical or wireless) and vise versa.

- There are two types of frame transmission: **asynchronous transmission** and **synchronous transmission**.
- Asynchronous transmission is character oriented, and each character has start and stop bits.
- There are two types of synchronous transmission: **character-oriented synchronization** and **bit-oriented synchronization**.
- In character-oriented transmission, a block of data is transferred over the network in the form of the ASCII code. Character-oriented transmission uses a SYN character for synchronization and information is transmitted character by character.
- Bit-oriented synchronization uses bit-oriented transmission. Information is transmitted bit by bit.
- The **IEEE 802 Committee** developed the standard for the Physical layer and Data Link layer.
- The IEEE 802 divides the Data Link layer into two sublayers: **Logical Link Controls (LLC)** and **Media Access Control (MAC)**.
- All IEEE 802.X use LLC frame format.
- The **IEEE 802.2** is a standard for the LLC.
- The **IEEE 802.3** is a standard for the Ethernet.
- The **IEEE 802.4** is a standard for the Token Bus.
- The **IEEE 802.5** is a standard for the Token Ring.
- The **IEEE 802.6** is a standard for MAN.
- The **IEEE 802.11** is a standard for wireless communication.
- The **IEEE 802.12** is standard for 100 VG–AnyLAN.
- The **IEEE 802.14** is standard for cable modem.

Key Terms

Application Layer
Asynchronous Transmission
Automatic Repeat Request (ARQ)
Bit-Oriented Synchronization
Character-Oriented Synchronization
Combined Station
Communication Protocol
Continuous ARQ
Data Link Layer
IEEE 802 Committee

International Standards Organization (ISO)
Logical Link Control (LLC)
Media Access Control (MAC)
Network Layer
Open System Interconnection Model (OSI)
Physical Layer
Presentation Layer
Primary Station

Secondary Station	Standards Organizations
Selective Reject ARQ	Stop-and-Wait ARQ
Session Layer	Synchronous Transmission
Sliding Window Method	Transport Layer

Review Questions

● **Multiple Choice Questions**

1. IEEE developed the _____ standard for LAN.
   - a. IEEE 802
   - b. RS-232
   - c. OSI model
   - d. All of the above

2. _____ defines standards for telecommunications.
   - a. IEEE
   - b. ITU
   - c. EIA
   - d. ISO

3. _____ defines standards for programming languages.
   - a. IEEE
   - b. ISO
   - c. ANSI
   - d. IEFT

4. _____ protocol is used on the Internet.
   - a. TPC/IP
   - b. X.25
   - c. IPX/SPX
   - d. NWLink

5. Microsoft's version of IPX/SPX is called _____.
   - a. NetBEUI
   - b. TCP/IP
   - c. NWLink
   - d. TCP/IP

6. OSI contains _____ layers.
   - a. 4
   - b. 3
   - c. 7
   - d. 6

7. A _____ layer establishes a connection.
   - a. Network
   - b. Physical
   - c. Data Link
   - d. Application

8. In character-oriented transmission, information is transmitted _____.
   - a. bit by bit
   - b. character by character
   - c. byte by byte
   - d. word by word

9. The _____ layer defines the format of the frame.
   a. Logical Link Control
   b. MAC layer
   c. Network layer
   d. Physical layer

10. Which layer of the OSI model is responsible for forming a frame? _____
    a. Data Link
    b. Transport
    c. Session
    d. Physical

11. Which layer of the OSI model performs encryption? _____
    a. Session layer
    b. Presentation layer
    c. Data Link layer
    d. Transport layer

12. The function of a network layer is _____.
    a. error detection
    b. routing
    c. to set up a session
    d. encryption

13. Which layer of the OSI model converts electrical signals to bits? _____
    a. Physical layer
    b. Data Link layer
    c. Network layer
    d. Application layer

14. Which layer determines the route for packets transmitted from source to destination? _____
    a. Data Link layer
    b. Network layer
    c. Transport layer
    d. Physical layer

15. Which of the following standards applies to Logical Link Control? _____
    a. IEEE 802.3
    b. IEEE 802.4
    c. IEEE 802.2
    d. IEEE 802.5

16. Which of the following standards applies to token ring? _____
    a. IEEE 802.3
    b. IEEE 802.5
    c. IEEE 802.2
    d. IEEE 802.4

17. What type of protocol is HDLC? _____
    a. Character-oriented
    b. Bit-oriented
    c. Character- and bit-oriented
    d. Word-oriented

● Short Answer Questions

1. Explain why it is necessary to have a standard for networking products.
2. List five standards organizations which are developing standards for networks and computers.
3. Define an open system.
4. Show an OSI model for a network.
5. List the function of the Application layer.
6. Explain the function of the Presentation layer.
7. Define the functions of the Session layer.
8. Explain the functions of the Transport layer.
9. List the functions of the Network layer.
10. Explain the functions of the Data Link layer.
11. Define the functions of the Physical layer.
12. IEEE 802 subdivides the Data Link layer into _____ sublayers: _____ and _____.
13. Show the frame format of IEEE 802.2.
14. Explain the function of MAC layer.
15. What is IEEE 802.3?
16. What is IEEE 802.5?
17. What are the frame transmission methods?
18. Explain asynchronous transmission method.
19. What does SDLC stand for?
20. List synchronous transmission methods.
21. What is an application of HDLC?
22. Explain character-oriented synchronization.
23. Explain bit insertion in bit-oriented synchronization.
24. Explain the function of synchronization bits.
25. Name an organization that makes standards for cable and connectors.
26. What is the function of SYN, STX, and ETX characters in character-oriented transmission?
27. List four communication protocols.
28. Explain communication protocol.
29. Which layer performs error detection?

30. Which layer converts electrical signals to bits?

31. Which layer is responsible for the physical connection between the source and the destination?

32. How many layers are in OSI model?

33. Which layer performs login?

34. Which layer performs information routing?

35. Show the frame format of character synchronization.

36. Show the frame format of bit synchronization.

37. Show the binary values of SYN, STX, and ETX characters.

38. List three types of HDLC frame.

39. Show the transmitted frame after bit insertion for following frame:

011111110000000011111011111110

chapter 5

Communication Channels and Media

OBJECTIVES   After completing this chapter, you should be able to:

- List the types of communication media currently in use
- Distinguish between different types of unshielded twisted-pair (UTP) cable
- List the different types of coaxial cable and their applications
- Discuss the different types of fiber-optics cable and their usage
- Explain the operation of wireless transmission

INTRODUCTION   Transmission medium is a path between the transmitter and the receiver in a transmission system. The type of transmission medium is defined by the various characteristics of the digital signal, including the signal rate, data rate, and the bandwidth of channel. The bandwidth of a channel determines the range of frequencies that the channel can transmit.

There are three types of communications media currently in use:

Conductive, such as twisted-pair wire and coaxial cable

Optical cable

Wireless

5.1   Conductive Media

The most popular **conductive media** used in networking are: unshielded twisted-pair cable (UTP), shielded twisted-pair cable (STP), and coaxial cable.

Twisted-Pair Cable   Unshielded twisted-pair (UTP) cable is the least expensive transmission medium and is typically used for LANs. Electrical interference, such as external electromagnetic noise generated by nearby cables, can have a devastating affect on the

79

performance of UTP cable. One way of improving the effect of noise on UTP cable is to shield the cable with metallic braid. **Shielded twisted-pair (STP) cable** provides better performance but is more difficult to work with. Figures 5.1(a) and 5.1(b) show samples of UTP and STP cable.

FIGURE 5.1(a)
Unshielded
twisted-pair
(UTP) cable

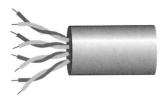

FIGURE 5.1(b)
Shielded
twisted-pair
(STP) cable

Unshielded twisted-pair cable is divided into categories CAT-1 through CAT-5 by the Electronic Industries Association (EIA). There are also proprietary enhancements to the CAT-5 specification that allow for even better performance over longer distances.

The EIA provides specifications for UTP cable, as shown in Table 5.1. These standards apply to four-pair UTP. UTP uses **RJ-45 and RJ-11 connectors**, as shown in Figure 5.2.

TABLE 5.1   UTP Specifications

Type of UTP	Performance	Application
CAT-1	None	None
CAT-2	1MHz	Telephone Wiring
CAT-3	16MHz	10BaseT, Token Ring 4 Mbps, ISDN low speed
CAT-4	20MHz	Token Ring 16
CAT-5	100MHz	100BaseT, 100 VG–AnyLAN, Token Ring 20 Mbps

FIGURE 5.2
RJ-45 and RJ-11
connectors

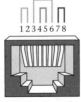

RJ-45 Female

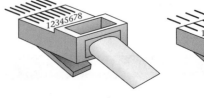

RJ-45 Male

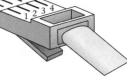

RJ-11

Coaxial Cable   **Coaxial cable** is used to transmit high-speed digital and analog signals over long distances. Figure 5.3 shows a coaxial cable that has an outer insulating cover made of polyvinyl chloride (PVC) or Teflon protecting the cable. Under the outer cover is

FIGURE 5.3
Coaxial cable

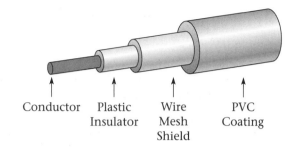

Conductor   Plastic   Wire   PVC
             Insulator   Mesh   Coating
                          Shield

a wire mesh shield, which provides excellent protection from external electrical noise. This shield is made of wire mesh or foil, or both. Under the shield is a plastic insulator, which isolates the center conductor from the shield.

The center conductor is a solid copper or aluminum wire that is shielded from external interference signals. Table 5.2 shows various types of coaxial cable, categorized by Radio Government (RG) rating. RG represents a set of specifications for cable such as the conductor diameter, thickness, and type of insulator. Coaxial cable uses a BNC connector, as shown in Figure 5.4.

TABLE 5.2 Various Types of Coaxial Cable and Their Applications

Cable Type	Impedance	Application
RG-59	75 ohms	Cable TV
RG-58	50 ohms, 5 mm in diameter	10Base2 or ThinNet
RG-11 and RG-8	50 ohms, 10 mm in diameter	10Base5 or ThickNet

FIGURE 5.4
BNC connector

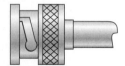

5.2 Fiber-Optic Cable

A **fiber-optic cable** is made of fiber which is covered by a buffer and a jacket. The fiber is composed of a core of thin glass or plastic covered by cladding which may also be glass or plastic. This fiber (the core and cladding) is then covered by a buffer to strengthen it. The buffer is finally covered by a plastic outer layer called the jacket which acts as a protective coating or shield. Figure 5.5 illustrates the structure of a fiber-optic cable.

To transmit information using optical fiber, the digital information is converted to light pulses by **light-emitting diodes (LED)** or **injected-laser diodes**

FIGURE 5.5
Fiber-optic cable

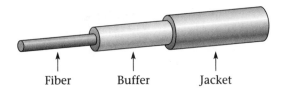

Fiber   Buffer   Jacket

(ILD) and sent through the fiber-optic cable. An LED is an electrical diode that generates a low power light. At the receiving end, a photodiode or a photo transistor is used to convert the light pulse signals back into electrical signals.

The advantages of fiber-optic cable are:

- Longer distance transmission due to reduced signal loss (attenuation)
- Greater bandwidth up to the Gigahertz range
- Immunity from any kind of noise or external interference such as electromagnetic signals'
- Smaller size
- Secure media

Some disadvantages of fiber-optic cables are:

- Network interface card and cabling is expensive
- Connection to the network is difficult

Types of Fiber-Optic Cable

In a fiber-optic cable, the angle of light reflection is directly dependent upon the diameter of the fiber. As the diameter increases, the light is reflected more and it takes more time to travel a given distance. There are two types of fiber-optics cable: Single-Mode Fiber (SMF) and Multimode Fiber (MMF).

1. **Single-Mode Fiber (SMF):** In **single-mode fiber**, only one light ray propagates through the fiber by bouncing to the wall core as shown in Figure 5.6. The core diameter of single-mode fiber is 9 micrometers and the wavelength of the light is 1.3 micrometers.

FIGURE 5.6
Single mode
fiber optics

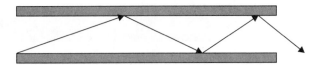

2. **Multimode Fiber (MMF):** In **multimode fiber** more than one light ray can propagate through the fiber, since each light ray propagates at different angles as shown in Figure 5.7. Multimode fiber has a core diameter larger than the wavelength of the light source being used. For multimode fiber, the core diameter ranges from 50 micrometers to 1000 micrometers, and the wave length of the light is about 1 micrometer.

FIGURE 5.7
Multimode step
index fiber optics
(MMF)

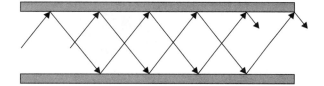

This means light can propagate through the fiber in many different ray paths, or modes. A single-mode fiber cable has a smaller diameter than multimode fiber cable.

There are two type of multimode fiber: multimode step index fiber and multimode graded index fiber.

A. **Multimode step index fiber:** This is a simple type of multimode fiber in which the index of refraction (the ability of the material to bend light) is the same all across the core of the fiber. Therefore, rays of light can propagate as shown in Figure 5.7. For step index fiber the bandwidth is typically 20 to 30 MHz over a distance of one kilometer.

B. **Multimode graded index fiber:** In multimode graded index fiber, the index of refraction across the core is gradually changed from maximum at the center to minimum near the edge. This type of fiber causes light to travel faster in the low index of refraction material than in the high-refraction material. Typical bandwidth for graded index fibers ranges from 100 MHz/Km to 1 GHz/km. Figure 5.8 also shows multimode graded index fiber-optic cable.

FIGURE 5.8
Multimode graded
index fiber

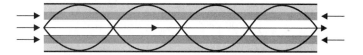

Fiber-Optic Connectors

There are three types of fiber-optic connectors used for networking. These connectors are listed below.

1. **Subscriber Channel Connector (SC):** The SC connector uses a push-pull locking system and is used for CATV and telephone connections, as shown in Figure 5.9.

2. **Straight Tip Connector (ST):** ST connector uses bayonet locking and is valued for its high reliability. It is also shown in Figure 5.9.

3. **MT-RJ Connector:** The MT-RJ connector is a duplex connector, as shown in Figure 5.10. The size of connector is equal to an RJ-45. The MT-RJ is a new connector for fiber-optic cable.

FIGURE 5.9
Fiber-optic ST and
SC connectors

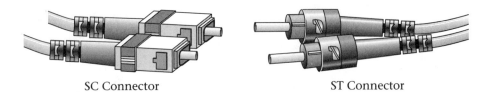

SC Connector   ST Connector

FIGURE 5.10
MT-RJ connector

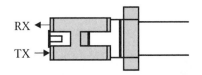

5.3   Wireless Transmission

Wireless transmission does not use any transmission media, such as a conductor or optical cable, to transmit and receive information. Microwave, radio, infrared light, and lasers are forms of wireless communication.

When electrons move, they generate electromagnetic waves. Wireless transmission uses these electromagnetic waves, which travel through space. Table 5.3 shows the electromagnetic wave spectrum and its applications.

TABLE 5.3   Electromagnetic Frequency Spectrum and Its Applications

Frequency Range	Name	Application
3–30 kHz	Very Low Frequency (VLF)	Telephone
30–300 kHz	Long Waves (LW)	Radio Frequency for Navigation
300–3000 kHz	Medium Waves (MW)	AM Radio Frequency
3–30 MHz	Short Wave (SW)	CB Radio Frequency
30–300 MHz	Very High Frequency (VHF)	TV and FM radio
300–3000 MHz	Ultra High Frequency (UHF)	TV
3–30 GHz	Super High Frequency (SHF)	Terrestrial and Satellite Microwave
30–300 GHz	Extreme High Frequency (EHF)	Experimental
>300 GHz	Infrared Light	TV Remote Control
	Lasers	Laser Surgery

Microwave Transmission   **Microwave transmission** uses electromagnetic waves to transmit information through the atmosphere. The range of microwave frequency is between 2 GHz and 40 GHz. Since any electromagnetic wave with a frequency of more than 100 MHz travels in a straight line, the type of transmission that microwaves provides is referred to as **Line-of-Sight** transmission.

There are two types of microwave communication. They are listed as follows:

1. **Terrestrial:** A typical system for transmitting and receiving microwave signals is a parabolic antenna or dish, as shown in Figure 5.11. Terrestrial microwave communication is typically used for long distance telecommunications and between buildings when wiring is impossible.

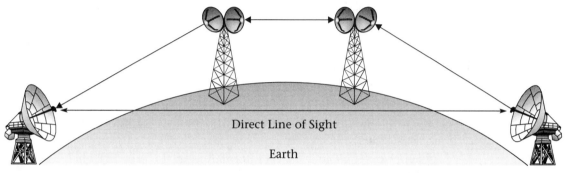

FIGURE 5.11 Terrestrial-based microwave system

2. **Satellite System:** A satellite system is used for communication between states and countries when it is not possible to use ground-based line of sight, as shown in Figure 5.12. The satellite is used to link two or more

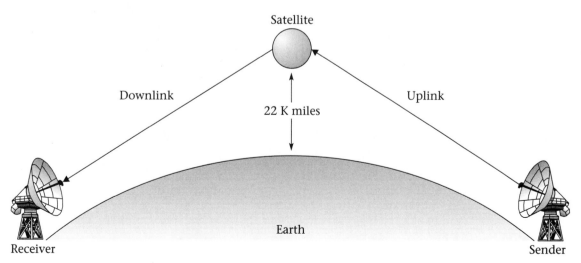

FIGURE 5.12 Satellite-based transmission system

ground stations together (earth station). The satellite receives the transmitted signals from the earth station (uplink) and retransmits the signals to another earth station (downlink). Two separate frequencies are used for uplink and downlink transmission.

Summary

- **Transmission media** is used to connect computers.
- The types of communication media currently in use are: **conductive, fiber-optic cable,** and **wireless.**
- The types of conductors used for networking are **Unshielded Twisted-Pair cable (UTP), Shielded Twisted-Pair cable (STP),** and **Coaxial cable.**
- UTP cable contains several pair of wires and is divided by EIA into categories such as CAT-1, CAT-5, and CAT-6.
- Unshielded twisted-pair (UTP) cable uses **RJ-45** or **RJ-11** connectors.
- Coaxial cable is used for transmitting high-speed information over relatively long distances.
- **Fiber-optic** cable transfers information in the form of light.
- The **Light-Emitting Diode (LED)** and **Injected Laser Diode (ILD)** are used to convert a digital signal to an optical signal for transmitting information over optical cable.
- There are two types of optical cable: **Single-Mode Fiber (SMF)** and **Multimode Fiber (MMF).**
- Single-mode fiber uses only one ray of the light source.
- In **MMF step index**, the index of refraction is the same all across the core of the fiber and a ray of light makes a sharp refraction.
- In **MMF grade index**, the index of refraction changes from maximum at the center to minimum near the edge of the core, causing the light to bend in a curved shape.
- **Wireless transmission** uses microwave, radio, or infrared light signals to transmit information.
- **Very High Frequency (VHF)** and **Ultra High Frequency (UHF)** use line of sight to transmit information.
- **Microwave** uses radio waves in the range of 1 GH to 23 GHz frequency.
- There are two types of microwave communication: **Terrestrial** and **Satellite System.**

Key Terms

Coaxial Cable

Conductive Media

Fiber-Optic Cable

Injected-Laser Diodes (ILD)

Light-Emitting Diodes (LED)

Line-of-Sight Transmission

Microwave Transmission

MT-RJ Connector

Multimode Fiber (MMF)

Multimode Graded Index Fiber

Multimode Step Index Fiber

RJ-45 and RJ-11 Connectors

Shielded Twisted-Pair Cable (STP)

Single-Mode Fiber (SMF)

Unshielded Twisted-Pair Cable (UTP)

Wireless Transmission

Review Questions

● Multiple Choice Questions

1. A _____ cable is the least expensive transmission media.
   - a. UTP
   - b. STP
   - c. fiber-optic
   - d. coaxial

2. _____ cable is used to transmit high-speed analog signals.
   - a. UTP and coaxial
   - b. STP and UTP
   - c. Coaxial and fiber-optic
   - d. Fiber-optic and STP

3. _____ connector(s) is/are used in fiber-optic cable.
   - a. SC and BNC
   - b. ST and SC
   - c. RJ-11 and ST
   - d. BNC and SC

4. _____ transmission does not use any transmission medium.
   - a. WAN
   - b. LAN
   - c. Wireless
   - d. Internet

5. Wireless transmission uses _____ waves.
   - a. optical true
   - b. electrical
   - c. electromagnetic
   - d. digital

6. Which of the following UTP cables is suitable for a data rate of 100 Mbps? _____
   - a. CAT-2
   - b. CAT-4
   - c. CAT-3
   - d. CAT-5

7. Which of the following transmission media are used for high-speed transmission? _____
   a. Coaxial cable and fiber-optic cable
   b. Fiber-optic cable and UTP CAT-2 cable
   c. Microwave and fiber-optic cable
   d. UTP CAT-2 cable and microwave cable

8. What type of fiber-optic cable is used for long-distance transmission? _____
   a. Multimode graded index   c. Multimode step index
   b. Single-mode   d. STP

● **Short Answer Questions**

1. List the major communication media.
2. What does UTP stand for?
3. What does STP stand for?
4. Which organization defines standards for cables?
5. What is the performance of a CAT-5 UTP?
6. What type of light source is used in fiber-optic cable?
7. What are advantages of fiber-optic cable over conductive cable?
8. What are the types of fiber-optic cable?
9. What does SMF stand for?
10. What does MMF stand for?
11. What is the application of single-mode fiber?
12. What is the application of multimode fiber?
13. What are the types of microwave communication?
14. What is the range of microwave frequency?
15. What are the signal sources for optical cable?
16. What are advantages of optical cable over coaxial cable?
17. What is advantage of STP cable over UTP cable?

chapter 6

Multiplexers and Switching Concepts

OBJECTIVES   After completing this chapter, you should be able to:

- Explain the operation of a multiplexer and a demultiplexer
- List the types of multiplexers and their operations
- Discuss how a telephone system operates
- Tell how pulse code modulation is used to convert voice to digital signals
- Explain T1 Link technology, including how to calculate its data rate
- Discuss switching concepts
- List the types of switching methods

INTRODUCTION   Long-distance transmission lines are expensive, so a method to allow several devices to share one transmission line is necessary to defray the cost of wiring. The multiplexer provides a solution to this problem. Figure 6.1 shows four terminals sharing one transmission line to send information to the host computer by using a multiplexer instead of four transmission lines.

FIGURE 6.1
Application of
multiplexer

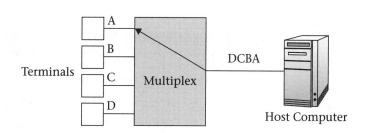

Terminals

A
B
C
D

Multiplex

DCBA

Host Computer

A **multiplexer** is a device which combines several low-speed data channels and transmits all of the data on a single high-speed channel. A common application of multiplexing is long-distance communication using high-speed point-to-point links for transferring large quantities of voice signals and data between users. Figure 6.2 shows the basic architecture of a multiplexer. A multiplexer that has N inputs and one output is called an N-to-1 Multiplexer. Figure 6.2 shows a 4-to-1 Multiplexer. The internal switch selects one input line at a time and transfers the input to the output. When the switch is in position A, it transfers input A to the output; then the switch moves to position B and transfer input B to the output. This process continues until the switch moves to position D and transfers input D to the output. After this function is complete, the switch starts over from A input.

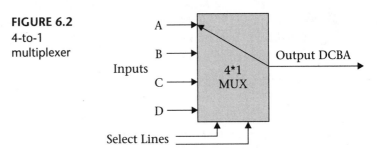

FIGURE 6.2
4-to-1
multiplexer

The opposite of a multiplexer is a **demultiplexer** (DMUX), shown in Figure 6.3. The switch moves to send each input to the appropriate output. DMUX has one input and N outputs—this is called a 1-to-N demultiplexer. When the switch is in position 0, it transfers A to output port 0, then moves to output port 1 and transfers B to this port. This process continues until the switch moves to output port 3 and transfers D to port 3. One cycle is complete and the transfer of data starts over from port 0.

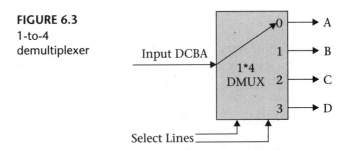

FIGURE 6.3
1-to-4
demultiplexer

6.1 Types of Multiplexers

There are four types of multiplexing. These methods are as follows:

1. Time Division Multiplexing (TDM)
2. Frequency Division Multiplexing (FDM)

3. Statistical Packet Multiplexing (SPM)

4. Fast Packet Multiplexing (FPM)

Time Division Multiplexing

In **Time Division Multiplexing (TDM)**, multiple digital signals can be carried on a single transmission path by interleaving each input of the multiplexer. A TDM operates on preassigned equal time slot to each input. It divides the bandwidth of the multiplexer's output into fixed segments. Each input to the MUX is given a fixed unit of time. First, information from input one is transmitted, then input from number two, and so on in a regular sequence, as shown in Figure 6.4.

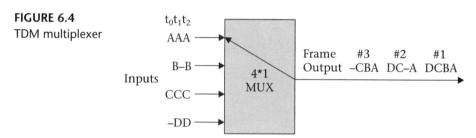

FIGURE 6.4
TDM multiplexer

One disadvantage of a TDM is that the bandwidth of this input is not available to other inputs when the input to the TDM is inactive. In Figure 6.4, the inputs to B and D are inactive at times t_1 and t_2, and the outputs in frame #2 and #3 have idle times. Another disadvantage of a TDM is that it is not able to change the bandwidth of the input dynamically, and therefore cannot transport a combination of voice, fax, and data.

Frequency Division Multiplexing

Frequency Division Multiplexing (FDM) divides the bandwidth of a transmission line into channels, and each channel can transmit specific information. Figure 6.5 shows the multiplexing of several TV channels using FDM. The bandwidth of coaxial cable is 500 MHz and can carry 80 TV channels. Each TV channel is assigned a different frequency, each using 6 MHz of bandwidth. Therefore, FDM combines several signals for transmission on a single transmission line.

Statistical Packet Multiplexing

Statistical Packet Multiplexing (SPM) dynamically allocates bandwidth to the active input channels, resulting in very efficient bandwidth utilization. In SPM, an idle channel does not receive any time allocation, as shown in Figure 6.6. SPM uses a store-and-forward mechanism in order to detect and correct any error from incoming packets. SPM will not allocate a time slot to any idle input.

Fast Packet Multiplexing

Fast Packet Multiplexing (FPM) uses the same method as SPM and has the ability to assign maximum bandwidth to any input needed. FPM does not use a store-and-forward mechanism, and therefore cannot perform error detection and correction. FPM will forward a packet before it has been completely received by the multiplexer.

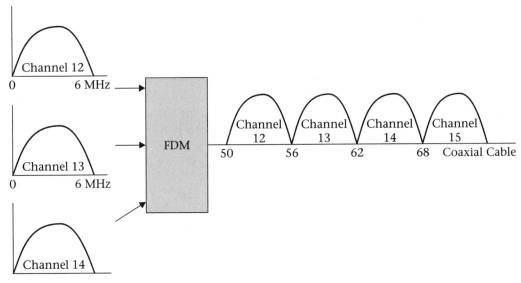

FIGURE 6.5 Frequency division multiplexing (FDM)

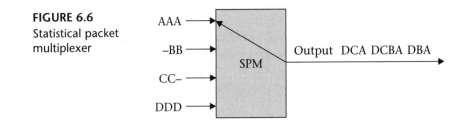

FIGURE 6.6
Statistical packet
multiplexer

6.2 Telephone System Operation

The current telephone system transmits information in analog signal from a telephone set to the Central Office (CO). At the CO, the analog signal is converted to a digital signal, which is transferred to the next central office. The voice is transmitted in the form of digital signals between the central offices, as shown in Figure 6.7. This digital signal is then converted to an analog signal and transmitted to the user. The method of conversion from analog to digital is called **Pulse Code Modulation (PCM)**.

6.3 Digitizing the Voice

Voice is an analog signal. In the central office of telephone company it is digitized by a device called a **codec** (coder-decoder). The function of a codec is to

FIGURE 6.7
Telephone system architecture

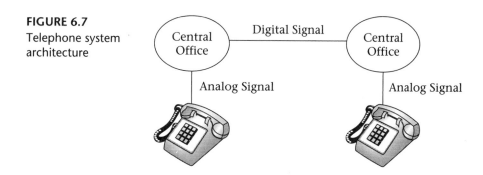

digitize the voice signal and to convert an already digitized signal to an analog. According to the Nyquist theorem, in order to convert voice into a digital signal, the analog signal must be sampled at the rate of two times the highest frequency of the voice signal. The highest frequency of human is 4000 Hz. Therefore, a codec samples the human voice at the rate of 8000 samples per second, as shown in Figure 6.8. This method is called **Pulse Amplitude Modulation (PAM)**.

FIGURE 6.8
Analog signal and pulse amplitude modulation (PAM)

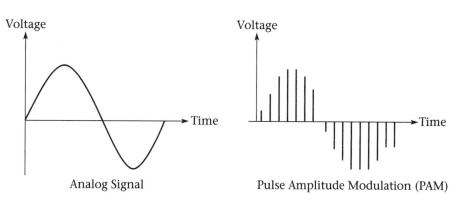

Each PAM sample is represented by eight bits. In Figure 6.9 it is represented by four bits. Remember, this method of converting voice to digital signal is called pulse code modulation (PCM). Since voice is digitized at the rate of 8000 samples per second and each sample is represented by eight bits, the data rate of the human voice is 8000 * 8 = 64 Kbps.

6.4 T1 Link

Long-distance carriers use TDM to transmit voice signals over high-speed links. One of the applications of TDM is the T1 link. A **T1 link** carries a level-1 digital signal (DS-1). A DS-1 is generated by multiplexing 24 voice digital signals (Digital Signal level-0 or DS-0), as shown in Figure 6.10. Pulse code

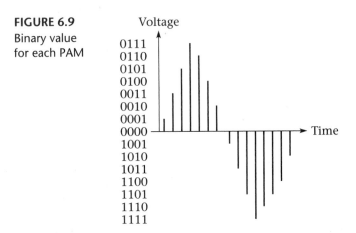

FIGURE 6.9
Binary value
for each PAM

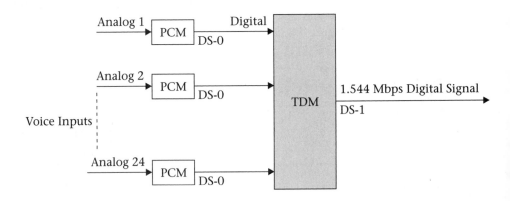

FIGURE 6.10
Architecture
of a T1 link

modulation is used to convert each analog signal to a digital signal. Each frame is made of 24 * 8 bits = 192 bits, with one extra bit added to separate each frame, making each frame 193 bits. Each frame represents 1/8000 of a second. Therefore, the data rate of a T1 link is 193 * 8000 = 1.544 Mbps.

Table 6.1 shows the TDM carrier standards for North America. Look at the table and you will see that DS-2 can carry 96 voice channels with 168 Kbps overhead; therefore, the data rate for DS-2 is 6.312 Mbps (96 * 64 Kbps + 168 Kbps overhead). Figure 6.11 shows the DS-1 frame format, the 1- bit gap is used to separate each frame.

6.5 Switching Concepts

A communication network that has more than two computers must establish links between the computers in order for them to be able to communicate with each other. One way to connect these computers is via a fully-connected network, shown in Figure 6.12.

TABLE 6.1 TDM Carrier Standard for North America

Frame Format	Line	Number of Voice Channels	Data Rates (Mbps)
DS-1	T1	24	1.544
DS-1C	T-1C	48	3.152
DS-2	T2	96	6.312
DS-3	T3	672	44.736
DS-4	T4	4032	274.176

FIGURE 6.11
DS-1 frame
format

1 bit	Byte #24	Byte #23		Byte #2	Byte #1

FIGURE 6.12
Fully-connected
network

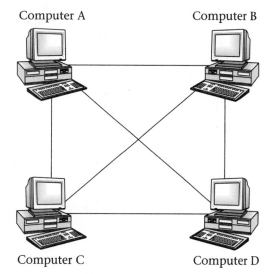

Computer A   Computer B

Computer C   Computer D

The advantage of this method is that all stations can communicate with each other. The disadvantage is that a large number of connections are required, especially when the number of stations is greater than four and they are some distance from each other. To overcome the above disadvantage, a device called a switch is used to connect stations, as shown in Figure 6.13.

The following methods are used to overcome the multipoint connection deficiency:

- Circuit Switching
- Message Switching

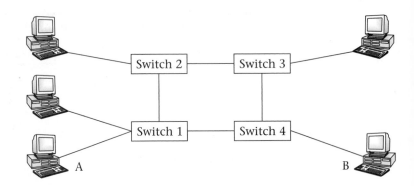

FIGURE 6.13
Stations are connected through switches

- Packet Switching
- Cell Switching

Circuit Switching   In **circuit switching**, also called a connection-oriented circuit, a physical connection must be established for the duration of the transmission (such as in a telephone system). The application of circuit switching is for real time communication. By dialing a telephone system, a connection is established and then communication begins, disconnecting at the end of the communication. Figure 6.14 shows circuit switching with multiple stations.

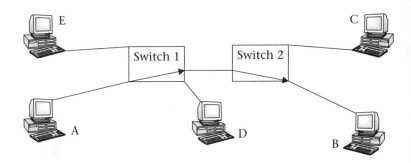

FIGURE 6.14
Circuit switching

Advantage of Circuit Switching   Circuit switching is used for real time communication. There is no delay or congestion in the communication link because a physical connection exists between the source and the destination.

Disadvantage of Circuit Switching   In circuit switching, only two stations use a communication link at the same time. Therefore, it is not cost effective. For example: In Figure 6.14, if stations A and B are communicating with each other, station D cannot communicate with station C. Station D must wait until A and B finish their communication, then D may start communicating with station C. In addition, if two stations such as A and B want to make a connection with C at the same time, a contention will occur and both stations must wait.

Message Switching In message switching, station A sends its message to the switch, the switch stores that message, and then forwards it to the destination. The disadvantage of message switching is that the switch needs to have a large buffer to store all of the incoming messages from other links.

Packet Switching Figure 6.15 shows a network with several switches. Source A has a message and wants to transfer it to destination B. Source A divides the message into packets and sends each packet by a different route. This process is known as **Packet Switching**. Each packet goes to the switch, which stores the packet and looks at the routing table inside the switch to find the next switch or destination. The packets may take different routes and be received at the destination out of order. To prevent mistakes in reassembling the packets, each packet has been given a sequence number that will be used by the destination to put the packets back in order.

FIGURE 6.15
Packet switching and virtual circuit

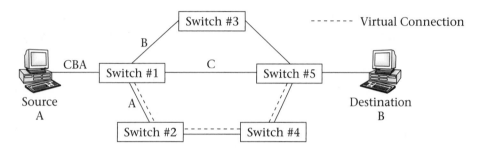

In Figure 6.15 the source divides the message into three packets: C, A, and B. The source transmits packet A to switch #1. Switch #1 stores packet A, looks at the congestion on all outgoing links, and finds that the link to switch #2 is the least congested. Switch #1 then sends packet A to switch #2. Switch #2 stores the packet, finds out from its routing table that packet A must go to Switch #4, and forwards packet A to switch #5. Packets B and C take different routes; therefore, the packets may be received out of order. The destination uses the sequence numbers of the packets to put them in proper order. The Internet uses packet switching. This type of service is also called a connectionless-oriented circuit.

Virtual Circuit Virtual circuits (a type of packet switching) operate on the same concept as packet switching, but the routing of the packets is specified before transmission. As seen in Figure 6.15, the source specifies the route, which is represented by dotted lines. Therefore all the packets from source A go via the dotted line. By using this method, all packets will be received at the destination in the proper order.

Summary

- A **multiplexer** is used to share a media among several users.
- In **Time Division Multiplexing (TDM)**, the users will each be assigned equal time to use the digital channel.

- **Frequency Division Multiplexing (FDM)** is used for analog transmission, with the bandwidth of the analog channel divided into smaller channels.
- In **Statistical TDM** the bandwidth is dynamically allocated to active users.
- The **Pulse Code Modulation (PCM)** method is used in the central switch to convert the human voice to digital signals.
- The bandwidth of human voice is 4000 Hz and it is digitized at the rate of 64 Kbps.
- A **T1 Link** is a special digital transmission line that has 24 inputs (each input 64 Kbps) of voice channels and one output with a data rate of 1.544 Mbps.
- There are three types of switching used in networking: Circuit Switching, Packet Switching and Cell Switching.
- A message is divided into pieces. Each piece is called a **packet**.
- **Packet Switching** treats each packet of a message separately.
- In **Circuit Switching** a physical connection must be established between the source and destination before transmitting information.
- **Virtual Circuit** is a type of packet switching, In virtual circuit, all packets of a message are transmitted in order along one path called a virtual path.

Key Terms

Circuit Switching	Packet Switching
Codec	Pulse Amplitude Modulation (PAM)
Demultiplexer	Pulse Code Modulation (PCM)
Fast Packet Multiplexing (FPM)	Statistical Packet Multiplexing (SPM)
Frequency Division Multiplexing (FDM)	T1 Link
	Time Division Multiplexing (TDM)
Multiplexer	Virtual Circuits

Review Questions

- **Multiple Choice Questions**

    1. Several devices can share one transmission line by using a _____.
        - a. multiplexer
        - b. demultiplexer
        - c. BUS
        - d. CPU

    2. _____ divides the bandwidth of a transmission line into channels.
        - a. TMD
        - b. FDM
        - c. SPM
        - d. FSPM

3. _____ dynamically allocates bandwidth to active inputs.
   a. TDM and FDM
   c. FPM and TDM
   b. SPM and FPM
   d. FDM and SPM

4. The highest frequency of a human voice is _____.
   a. 4000 Hz
   c. 40 Hz
   b. 400 Hz
   d. 4 Hz

5. One of the applications of TDM is a _____ link.
   a. DSL
   c. cable modem
   b. T1
   d. LAN

6. _____ is a type of packet switching.
   a. Circuit switching
   c. Virtual circuit
   b. Packet switching
   d. Message switching

7. Pulse code modulation is used to convert _____.
   a. digital to analog
   c. digital to digital
   b. analog to digital
   d. none of the above

8. Which of the following switching methods delivers packets in order? _____
   a. Packet switching and Virtual circuit
   b. Virtual circuit and Circuit switching
   c. Circuit switching and Packet switching
   d. none of the above

9. The bandwidth of a telephone system is _____.
   a. 3 kHz
   c. 8 kHz
   b. 4 kHz
   d. 1 kHz

● **Short Answer Questions**
   1. Show an 8-to-1 MUX and a 1-to-8 DMUX.
   2. List the types of multiplexers.
   3. Explain TDM operation.
   4. Describe a statistical packet multiplexer.
   5. What is the type of signal used between two central switches of a telephone system?
   6. What is the function of a codec?
   7. What does PCM stand for and what is its application?

8. Why must the human voice be sampled at the rate of 8000 samples per second?

9. What type of multiplexer is used in a T1 link?

10. What is the data rate of the human voice and why?

11. What is the data rate of a T1 link?

12. What is difference between DS-1 and a T1 link?

13. What is the data rate of a T3 link?

14. How many voice channels can be carried by a T3 Link?

15. Explain the following switching operations:
    a. Circuit Switching
    b. Message Switching
    c. Packet Switching
    d. Virtual Circuit

16. How many inputs, outputs and select lines does a 1 * 8 demultiplexer have?

17. How many inputs, outputs and select lines does a 16 * 1 multiplexer have?

18. Show the frame format of a T1 link.

19. What data rate of a T1 link is 1.54 Mbps?

20. The following inputs are connected to a 4 * 1 statistical multiplexer. Show the outputs of the multiplexer.
    a. Input #1   A-A-A
    b. Input #2   BBB-B
    c. Input #3   - CC- -
    d. Input #4   DD-D

21. What is the sampling rate of a signal with the highest frequency of 1000 Hz?

chapter

Modem, DSL, Cable Modem, and ISDN

OBJECTIVES

After completing this chapter, you should be able to:

- Discuss modem operation and the methods of signal modulation
- Explain the operation of the 56 K modem
- Understand the technology of Digital Subscriber Line (DSL) and xDSL
- Explain cable modem technology
- Have a clear understanding of ISDN modem technology

INTRODUCTION

In order for two computers to communicate with each other, a link between them is required. If these computers are some distance from each other, it is not cost effective to use a cable and link them together. A cheaper alternative is to use a telephone or cable TV line to provide the link. The device that enables users to establish a link between computers using a telephone line is called a modem. The types of modem that use telephone lines are the dial-up modem, the Integrated Digital Services Network modem (ISDN modem), and the Digital Subscriber Line modem (DSL modem).

Currently, about 63 million households have cable TV service. The same wire that bring TV signals to your house is a cable that can also provide Internet access with a speed 100 times faster than the dial-up modem. The device that enables computers to access the Internet by cable TV lines is called a cable modem.

7.1   Modem

To link computers for communication over traditional telephone wires, a modem must be used, as shown in Figure 7.1.

FIGURE 7.1
Connection of two computers using a modem

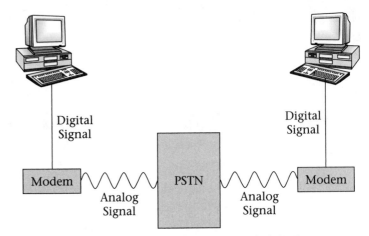

Digital Signal

Digital Signal

Modem ⟋⟍⟋⟍ PSTN ⟋⟍⟋⟍ Modem

Analog Signal

Analog Signal

Public Switch Telephone Network (PSTN)

A traditional telephone network operates with analog signals, whereas computers work with digital signals. Therefore a device is required to convert the computer's digital signal to an analog signal compatible with the phone line (**modulation**). This device must also convert the incoming analog signal from phone line to a digital signal (**demodulation**). Such a device is called a **modem**; its name is derived from this process of **mo**dulation/**dem**odulation.

A modem is also known as Data Circuit Terminating Equipment (DCE), which is used to connect a computer or data terminal to a network. Logically, a PC or data terminal is called Data Terminal Equipment (DTE).

There are three types of modems. An **internal modem** is an expansion card that plugs into an ISA or PCI bus inside the computer. It is connected to a phone line by an RJ-11 connector. There is also an **external modem**, available. Its circuitry is housed in a separate casing and it typically uses a DB-9 connector to attach to one of the computer's serial ports. The third type is used in laptop and notebook computers and consists of a PC card that houses the entire circuitry for the modem.

A modem's transmission speed can be represented by either data rate or baud rate. The **data rate** is the number of bits which a modem can transmit in one second. The **baud rate** is the number of signals which a modem can transmit in one second.

Modulation Methods

The carrier signal on a telephone line has a bandwidth of 4000 Hz. Figure 7.2 shows one cycle of telephone carrier signals. The following types of modulation are used to convert digital signals to analog signals:

- Amplitude Shift Keying(ASK)
- Frequency Shift Keying (FSK)
- Phase Shift Keying (PSK)
- Quadrature Amplitude Modulation (QAM)

FIGURE 7.2
Telephone carrier
signal

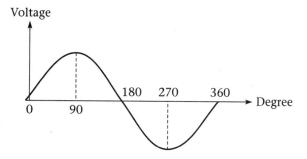

Voltage

Amplitude Shift Keying   In **Amplitude Shift Keying (ASK)**, the amplitude of the signal changes. This also may be referred to as Amplitude Modulation (AM). The receiver recognizes these modulation changes as voltage changes, as shown in Figure 7.3. The smaller amplitude is represented by *zero* and the larger amplitude is represented by *one*. Each cycle is represented by one bit, with the maximum bits per second determined by the speed of the carrier signal. In this case, the baud rate is equal to the number of bits per second.

FIGURE 7.3
Amplitude shift
keying (ASK)

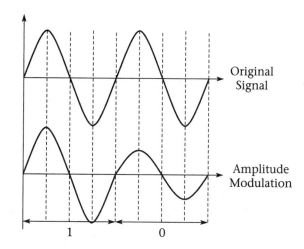

Original
Signal

Amplitude
Modulation

Frequency Shift Keying   With **Frequency Shift Keying (FSK)**, a *zero* is represented by no change to the frequency of the original signal, while a *one* is represented by a change to the frequency of the original signal. This is shown in Figure 7.4. Frequency modulation is a term often used in place of FSK.

Phase Shift Keying   Using the **Phase Shift Keying (PSK)** modulation method, the phase of the signal is changed to represent *ones* and *zeros*. Figure 7.5 shows a 90-degree phase shift. Figures 7.6(a), (b), and (c) show the original signals with a 90-degree shift, a 180-degree shift and a 270-degree shift, respectively.

FIGURE 7.4
Frequency shift
keying (FSK)

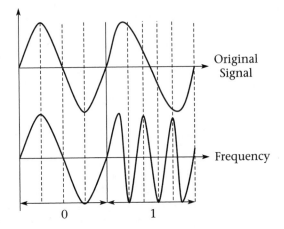

FIGURE 7.5
90-degree phase shift

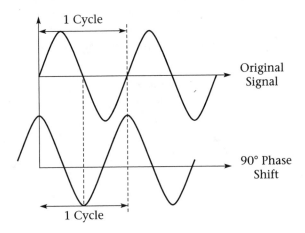

FIGURE 7.6
Phase shift for 90,
180, and 270
degrees

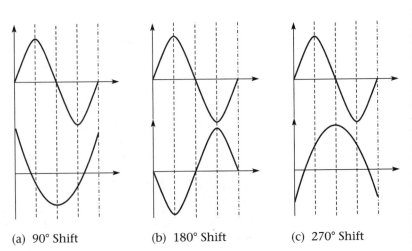

(a) 90° Shift   (b) 180° Shift   (c) 270° Shift

In Figure 7.6, note that the original signal can be represented with four different signals: no shift, 90° shift, 180° shift, and 270° shift. Therefore, each signal can be represented by a 2-bit binary number, as shown in Table 7.1 below.

The modem's speed using a 90-degree phase shift is 2 * 4000, which is equal to 8 Kbps. To increase the speed of the modem, the original signal can be shifted 45 degrees to generate eight distinct signals. Each signal is represented by three bits. Therefore, the speed of the modem is increased to 3 * 4000 which is equal to 12 Kbps.

The relation between phase and the binary representation of each phase can be plotted on a coordinate system called a **constellation diagram**. Figure 7.7 is a constellation diagram showing the four distinct signals of a 90-degree shift, with each signal represented by two bits. Figure 7.8 shows a constellation diagram using 45-degree shift and 3-bit representation (8-PSK).

FIGURE 7.7
Constellation
diagram
for Table 7.1

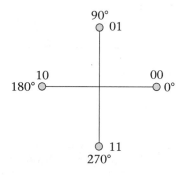

TABLE 7.1 Binary Values of Signals

Phase Shift	Binary Value
No Shift	00
90°	01
180°	10
270°	11

FIGURE 7.8
Constellation
diagram
for 8-PSK

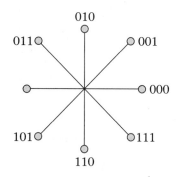

Bits	Phase Shift
000	0°
001	45°
010	90°
011	135°
100	180°
101	225°
110	270°
111	315°

Quadrature Amplitude Modulation One method of increasing the transmission speed of a modem is to combine PSK and ASK modulation. This hybrid modulation technique is called **Quadrature Amplitude Modulation (QAM)** and is shown in Figure 7.9. Here we see the combination of four phases and two amplitudes which generates eight different signals called 8-QAM. Table 7.2 shows the binary value of each signal and provides a constellation diagram for 8-QAM. See Figure 7.10. The data rate of the modem is 3 bits * 4 K = 12 Kbps.

FIGURE 7.9
8-QAM
modulation

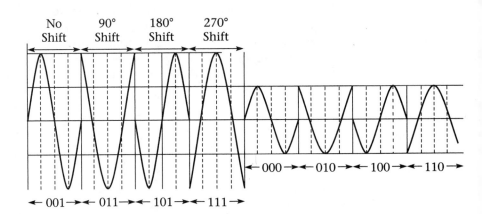

FIGURE 7.10
Constellation
diagram for
8-QAM

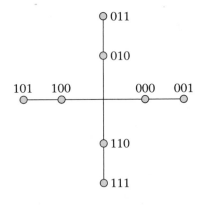

TABLE 7.2 Binary Value for 8-QAM

Shift	Amplitude	Binary Value
No	A1	000
No	A2	001
90°	A1	010
90°	A2	011
180°	A1	100
180°		
0		

Modem Standards The International Telecommunication Union (ITU) is responsible for developing standards for modems. Currently, most modem manufacturers produce modems with a transmission speed of up to 56 Kbps. Following are some modem standards and their respective speeds.

Name	Speed
V.90 or X2	56 Kbps receiving
	33.6 Kbps transmitting
V.36	48 Kbps
V.34	28.8 Kbps
V.33	14.4 Kbps
V.32	14.4 Kbps
V.26 bis*	1200, 2400 Kbps
V.22 bis	1200 Kbps

* bis means the modem has a switch and works with two different data rates

V.90 (56 Kbps) Modem The maximum theoretical data rate for a modem, as set by the ITU, is 33.6 Kbps. Several manufacturers have, however, developed 56 Kbps modems. The 56 K modem was designed for a one-end digital connection from the server to the Public Switch Telephone Network (PSTN); the subscriber side (the line that connects to the actual modem) remains analog. Figure 7.11 shows an application of a 56 K modem in which the server is connected to the PSTN without any modem and transmits information as a digital signal to the PSTN. At the central switch, digital information is converted to an analog signal and transmitted to the subscriber side. Only one side uses the modem, which reduces the signal/noise ratio caused by conversion and allows the server to transmit data at a rate of 56 Kbps. At the user side, the information is converted from digital to analog and transmitted to the central switch. These conversions produce noise and reduce the speed of the modem to 33.6 Kbps.

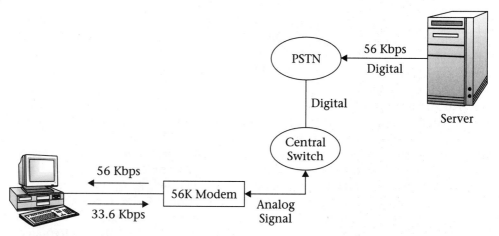

FIGURE 7.11   56 Kbps modem connection

7.2 Digital Subscriber Line

The **Digital Subscriber Line (DSL)** is the latest modem technology, using twisted-pair wires to deliver data and voice at speeds ranging from 64 Kbps to 50 Mbps. DSL uses current telephone wire (UTP) to transfer information at higher data rates than a modem. Currently, a modem transfers data at 56 Kbps, and networking technology transfers data at the rate of 10 to 1000 Mbps. Therefore, modems are becoming far too slow for transferring information across the Internet. DSL uses standard phone twisted-pair cable to transfer digital signals for data and analog signals with Plain Old Telephone Service (POTS). DSL is implemented using several different technologies called xDSL.

Asymmetrical DSL: ADSL supports voice and data simultaneously. The data rate from service provider to the user is 6 Mbps and 786 Kbps from the user to the service provider (telephone switch).

High Bit Rate DSL: HDSL supports data or voice, but not simultaneously, with a data rate of 768 Kbps.

Symmetrical DSL: SDSL supports voice and data simultaneously, with a data rate of 768 Kbps in both directions.

Very High Speed DSL: VDSL provides 25 to 50 Mbps to the user (downstream), and 1.5 to 3 Mbps from the user to the service provider (upstream).

Asymmetrical Digital Subscriber Line

The **Asymmetrical Digital Subscriber Line (ADSL)** is a new modem, which uses an existing twisted-pair telephone line to access the information superhighway for transferring information such as multimedia. ADSL transfers data at a higher rate **downstream** (from the telephone company switch) to the subscriber than **upstream** (from subscriber to telephone company switch). The upstream and downstream data rates are factors of the distance between the telephone company switch and the subscriber. The downstream data rate is from 1.5 to 8 Mbps and upstream data rate is from 16 Kbps to 640 Kbps, as shown in Figure 7.12. The advantage of ASDL is that it uses the present twisted-pair wire of telephone lines to transmit data in the range of 6 to 8 Mbps. It is important to note that ADSL does not affect the current telephone voice channel.

ADSL Modem Technology

ADSL uses Discrete Multi-Tone (DMT) encoding methods, which use QAM to divide the bandwidth of the channel into multiple subchannels, with each channel transmitting information using QAM modulation. The twisted-pair cable used in telephone wire has a frequency spectrum of 1.1 MHz. Figure 7.13 shows the frequency spectrum of ADSL. DMT uses the frequency spectrum from 26 kHz to 1.1 MHz for broadband data. For POTS it uses the frequency spectrum from 0 to 4 kHz.

The frequency spectrum from 26 kHz to 138 kHz is used for upstream transmission, and the frequency spectrum from 138 kHz to 1.1 MHz is used for downstream transmission, as shown in Figure 7.13. The frequency spectrum above 26

FIGURE 7.12
ADSL modem
connection

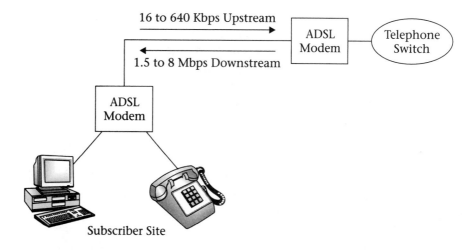

16 to 640 Kbps Upstream

1.5 to 8 Mbps Downstream

ADSL Modem

Telephone Switch

ADSL Modem

Subscriber Site

FIGURE 7.13
Frequency
spectrum of
ADSL

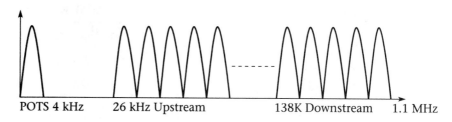

POTS 4 kHz   26 kHz Upstream   138K Downstream   1.1 MHz

kHz is divided into 249 independent subchannels, each containing 4.3 kHz bandwidth. Each subfrequency is an independent channel and has its own stream of signals. The lower 4-kHz channel is separated by an analog circuit and used in POTS; 25 channels are used for upstream transmissions, and 224 channels are used for downstream transmissions.

Figure 7.14 shows ADSL modem architecture. The function of the POTS filter is to separate the voice channel from data channel.

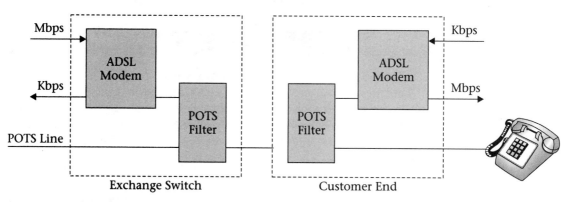

Mbps

Kbps

POTS Line

ADSL Modem

POTS Filter

Exchange Switch

Kbps

Mbps

ADSL Modem

POTS Filter

Customer End

FIGURE 7.14 ADSL modem architecture

Each subchannel can modulate from 0 to 15 bits per signal. This allows up to 60 Kbps per channel (15 * 4 kHz) and some implementation support of 16 bits per channel. Therefore, the data rate is calculated by:

Data Rate = Number of Channels * Number of Bits/Channel * Frequency of Channel

Using the above equation, the upstream and downstream data rate can be computed as follows:

Maximum Upstream Data Rate = 25 * 15 * 4.3 kHz = 1.6 Mbps and

Maximum Downstream Data Rate = 224 * 15 * 4.3 kHz = 14.4 Mbps

The data rate of an ADSL modem is a factor of the distance between the subscriber and the telephone switch. Table 7.3 shows the data rate vs. distance for an ADSL modem.

TABLE 7.3   Data Rate of ADSL Modem vs. Distance

Data Rate	Wire Gage	Distance
1.5–2 Mbps	24 AWG	5.5 km or 18000 ft
1.5–2 Mbps	26 AWG	4.6 km or 15000 ft
6.1 Mbps	24 AWG	3.7 km or 12000 ft
6.1 Mbps	26 AWG	2.7 km or 9000 ft

Rate Adaptive Asymmetric DSL   Rate Adaptive Asymmetric DSL (RADSL) offers downstream transmission of 7.0 Mbps and upstream transmission of 1.0 Mbps. The data rate of RADSL is dynamically set by the condition of the line.

Before transmitting information, RADSL makes an initial test to check the condition of the channels. Some of the channels may not be used due to the presence of strong noise. Table 7.4 shows xDSL and cable distance.

7.3   Cable Modem

The **Cable Modem** is another technology used for remote connection to the Internet. Residential access to the Internet is growing, and current modem technology can transfer data only at 56 Kbps. Local telephone companies also offer a service known as Basic Rate ISDN, which has a transmission rate of 128 Kbps. The cable modem offers high-speed access to the Internet using a media other than phone lines.

Cable TV System Architecture   Cable TV is designed to transmit broadband TV signals to homes using coaxial and fiber-optic cable. Figure 7.15 shows the full coaxial cable TV system architecture. As

TABLE 7.4   xDSL and Cable Distance

Technology	Cable Distance in Feet	Data Rate Downstream/Upstream
ADSL	3000	9 Mbps/1 Mbps
ADSL	5000	8.448 Mbps/1 Mbps
ADSL	9000	7 Mbps/1 Mbps
ADSL	12000	6.312 Mbps/640 Kbps
ADSL	18000	1.544 Mbps/16–64 Kbps
HDSL	5000	1.544 Mbps/1.544 Mbps
HDSL	12000	1.544 Mbps/1.544 Mbps
RADSL	3000	12Mbps/1 Mbps
RADSL	9000	7 Mbps/1 Mbps
RADSL	12000	6 Mbps/1 Mbps
RADSL	18000	1 Mbps/128 Kbps
UDSL	0–15000	2 Mbps/2 Mbps
UDSL	15000–18000	1 Mbps/1 Mbps
VDSL	1000	51.84 Mbps/2.3 Mbps
VDSL	3000	25.82 Mbps/2.3 Mbps
VSDL	4000	12.98 Mbps/1.6 Mbps

Source: Data Communication Magazine, April 1998

FIGURE 7.15
Full coaxial cable TV system architecture

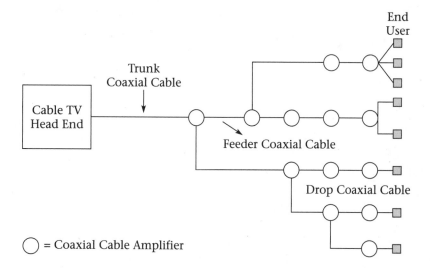

= Coaxial Cable Amplifier

shown in this diagram, cable TV uses Tree and Branch Bus Topology. The tree and branch cable is constructed of 75-ohm coaxial cable connected to the trunk cable.

The head end transmits TV signals over a **trunk cable** to a group of subscribers. The medium can be either coaxial or fiber-optic cable. The function of a **coaxial amplifier** is to amplify the signal, and it works in either direction. Feeder and drop cables are both coaxial cables. The **drop cable** is the part of a cable system that connects the subscriber to the feeder cable. **Feeder cables** are connected to a trunk cable to cover a large area. The maximum distance between head end and subscriber is 10 to 15 km. The maximum number of cascaded amplifiers is 35 and maximum number of connections is 125,000.

A TV signal transmits two types of frequencies: VHF (Very-High Frequency) and UHF (Ultra-High Frequency). The VHF channels use lower frequencies to generate stronger signals needed to transmit longer distances. The VHF channels start at channel 2 and end at channel 13, with a frequency up to 54 MHz. The UHF channels use the frequency range of 54 MHz to 216 MHz. UHF channels start at channel 14, with a frequency of 470 MHz. Each TV channel occupies 6 MHz of the TV radio frequency (RF) spectrum.

The bandwidth of coaxial cable is 500 MHz, with each TV channel requiring 6 MHz of bandwidth. The number of TV channels that can be transmitted is $(500 - 50)/6 = 75$ channels. In order to increase the bandwidth of cable TV, cable TV corporations use Hybrid Fiber Cable (HFC), which is combination of fiber-optic cable and coaxial cable, as shown in Figure 7.16. The bandwidth of a cable TV system using HFC cable is 750 MHz to 1GHz. Therefore, the number of channels computed by $(750 - 50)/6 = 110$ channels. The TV signal is transmitted to a fiber node using optical cable. The fiber node converts the optical signal to an electrical signal, and also converts electrical to optical. The coaxial amplifiers are two-way devices used to amplify the incoming signal. The maximum distance from head end to end user is 80 km and the maximum number of end users per fiber node connection is between 500 and 3000 (depending on the vendor). A channel between 5 MHz and 42 MHz is used to carry upstream signals (from subscriber to the head end).

FIGURE 7.16
HFC cable TV architecture

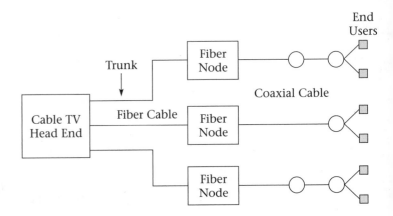

Cable Modem Technology

Figure 7.17 shows the components of a cable network consisting of a coaxial cable, **head end,** and a cable modem. The connection between cable modem and user is **10BaseT**. The user requires a 10BaseT NIC card to be able to use the cable modem. A cable modem can support more than one station by using a repeater, as shown in Figure 7.18.

FIGURE 7.17
Block diagram of a cable modem

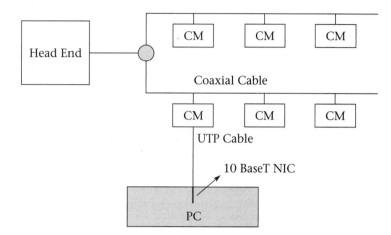

FIGURE 7.18
Connection of more than one station to a cable modem

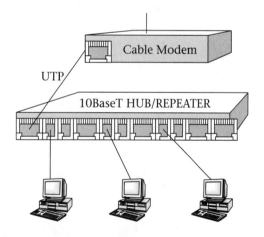

A cable modem uses 64-QAM or 256-QAM modulation techniques to transmit information from the head end to the cable modem (downstream transmission). If a cable modem uses 256 QAM, this means it transmits 8 bits per signal and each signal is transmitted at 6 MHz. Theoretically the data rate of a cable modem is:

8 * 6 MHz = 48 Mbps

By using 64-QAM modulation, the data rate becomes:

6 * 6 MHz = 36 Mbps

Upstream transmission (from cable modem to head end) uses a 600 kHz channel between 5 and 42 MHz. This low frequency is close to the CB radio frequency. The Quadrature Phase Shift Keying (QPSK) modulation method is used. In QPSK the signal is represented in four different forms such as: no shift, 90° shift, 180° shift and 270° shift. In this method each signal can be represented by two bits. The data rate of the cable modem for upstream transmission becomes:

2 * 600 kHz = 1200 Kbps

Downstream and upstream bandwidth are shared by 500 to 5000 cable modem subscribers. If 100 subscribers are sharing a 36-Mbps connection, each user will receive a data at rate of 360 Kbps. A cable modem provides a constant connection (like a LAN); it does not require any dialing. The cable modem head end communicates with the cable modem, and when the cable modem is commanded by cable modem head end, the modem will select an alternate channel for upstream transmission.

IEEE 802.14 Standard

A cable modem operates at the physical and data link layers of the OSI model. The **IEEE 802.14 standard** provides a network logical reference model for the media access control (MAC) and physical layer. The following are general requirements defined for cable by IEEE 802.14:

- Cable modems must support symmetrical and asymmetrical transmission in both directions
- Support of Operation, Administration and Maintenance (OAM) functions
- Support the maximum 80 km distance for transmission from the head end to the user
- Support a large number of users
- MAC layer should support multiple types of service, such as data, voice, and images
- MAC layer must support unicast, multicast, and broadcast service
- MAC layer should support fair arbitration for accessing the network

7.4 Integrated Services Digital Network

The **Integrated Services Digital Network (ISDN)** is a set of digital transmission standards which are used for end-to-end digital connectivity. ISDN supports voice and data. The integration of different services has become an ISDN hallmark. In the past, video, audio, voice, and data services required at least four separate networks. ISDN integrates all four over the same network. ISDN uses a digital signal which is less vulnerable to noise compared to the analog signal used by a modem.

ISDN brings the digital network to users and offers high-speed communication. There are two types of ISDN: **Narrowband ISDN (N-ISDN)** and **Broadband ISDN (B-ISDN)**.

The International Telecommunications Union (ITU) (formerly known as CCITT), has defined standards for ISDN that provide end-to-end digital connection to support a wide range of services including voice and non-voice transmission. ISDN offers digital transmission over existing telephone wiring as provided by telephone companies. ISDN offers Basic Rate Interface (BRI) and Primary Rate Interface (PRI).

Basic Rate Interface The **Basic Rate Interface (BRI)** is made up of two B-channels (bearer) and one D-channel. Therefore, the total rate is 2B + D. B-channels are 64 Kbps and can be used for voice and data communication. The D-channel is 16 Kbps and is used for call initialization and signaling connection. Figure 7.19 shows ISDN Basic Rate Interface.

FIGURE 7.19
ISDN basic rate interface

2B Voice or Data Channels
1D Signaling/Control Channels

Basic Rate Interface (BRI)

64 Kbps
16 Kbps

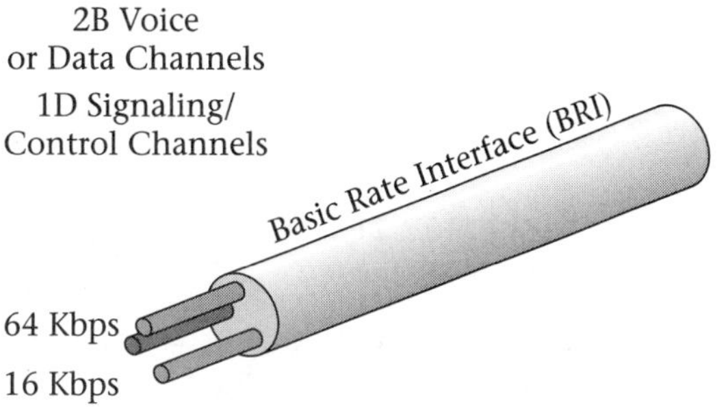

Application of BRI ISDN can carry multiple services—voice, video and data—on a single telephone line over existing twisted-pair copper wire. Remember, ISDN's BRI uses two 64 Kbps B-channels and one 16 Kbps D-channel. By combining two B-channels, the total data transmission rate is 128 Kbps. Figure 7.20 shows an application of BRI in ISDN.

Network Terminator Type 1 (NT1) and a power supply are required for every ISDN line. NT1 is a device that is physically connected to the ISDN line. A special terminal adapter can combine the two B-channels to create a 128 Kbps channel, which can then be connected to the computer.

The NT1 device works as multiplexer and demultiplexer. Figure 7.21 shows the frame format of ISDN Basic Rate Interface. The size of a frame is 48 bits.

Frame size = 4 * 8 + 4 + 12 bits overhead

Primary Rate Interface The **Primary Rate Interface (PRI)** in North America has 23 B-channels and one 64 K D-channel or 23B + D, having a total bandwidth of 1.544 Mbps. They are designed to replace T1 links. PRI in Europe uses 30 B-channels and one D-channel or 30B + D with total rate of 2.048 Mbps as shown in Figure 7.22.

FIGURE 7.20
Application of basic
rate interface

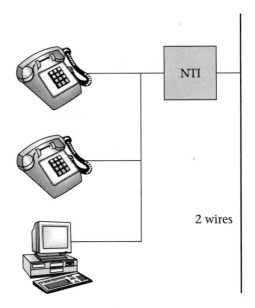

2 wires

FIGURE 7.21
ISDN BRI frame
format

8 bits	1 bit	8 bits	1 bit	8 bits	1 bit	8 bits	1 bit
B1	D	B2	D	B1	D	B2	D

FIGURE 7.22
ISDN primary
rate interface

23B (US, Japan) or 30B (Europe)
Voice or Data Channels

1D Signaling/
Contol Channels

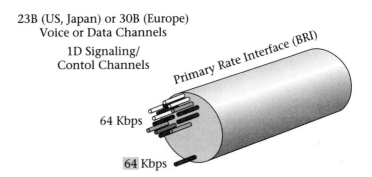

Application of PRI An application of the PRI is to connect two central switches together to use them as a T1 link, as shown in Figure 7.23. The devices which handle switching and multiplexing (such as PBX) are called Network Terminator Type 2 (NT2). ISDN Primary Rate Interface (PRI) can connect the customer directly by using an NT2 device, while ISDN BRI requires an NT1 device.

FIGURE 7.23
ISDN PRI
application

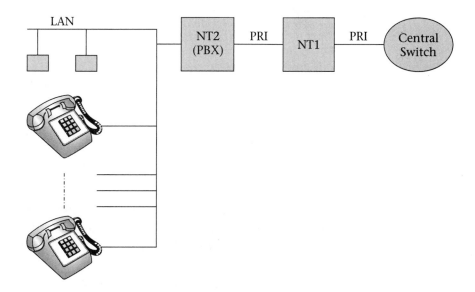

Advantages of ISDN   Since ISDN is a digital service, the distance of the user to the central office should not exceed 18,000 feet.

The following are some advantages of the ISDN over a modem:

- ISDN offers three separate channels, two B-channels and one D-channel.
- ISDN can transfer two times faster than a 56 K modem.
- ISDN uses a digital signal, which is less vulnerable to noise than an analog signal.
- ISDN offers clear voice conversations in comparison to the current telephone system.
- An ISDN modem takes two to four seconds to establish a connection, and an analog modem takes about 15 to 30 seconds to dial and make a connection.
- ISDN provides clear, quieter voice telephone service and easy-to-use call control features. Its Caller Identification features can screen incoming calls, much like today's Caller ID services.

Four types of modem technology have been presented: dial-up modem, DSL modem, cable modem. There are some features of each which you need to remember.

The maximum data rate for the dial-up modem is 56 Kbps. The maximum data rate for an ISDN modem is 128 Kbps, and the cable modem data rate is variable and depends on how many users are accessing the link at the same time. (Also, a cable TV link is shared with other users, making security and privacy an issue to take into account.)

The xDSL modem is the latest technology in that this modem uses the telephone lines and can transmit data at higher rates than an ISDN modem or dial-up modem. One of the biggest advantages of the xDSL modem over cable modem is improved security and privacy. The xDSL modem is more cost effective, since it will replace the expensive T1 link. Because it is cost effective, it can be used for connecting small libraries and schools to Internet.

Summary

- The function of a modem is to convert analog signals to digital and digital signals to analog.
- **Modulation** is used to convert a digital signal to analog.
- The modulation methods are: Amplitude Shift Keying (ASK), Frequency Shift Keying (FSK), Phase Shift Keying (PSK), and Quadrature Amplitude Modulation (QAM).
- **Amplitude Shift Keying (ASK)** changes the amplitude of carrier signals in order to represent a digital signal.
- **Frequency Shift Keying (FSK)** changes the frequency of carrier signals in order to represent a digital signal.
- **Phase Shift Keying (PSK)** changes the phase of carrier signals.
- **Quadrature Amplitude Modulation (QAM)** is a combination of ASK and PSK, and is used in high-speed modems.
- **Baud rate** is the number of signals per second that a modem can transmit.
- **Data rate** is the number of bits per second that a modem can transmit.
- A **Digital Subscriber Line** (DSL) is a type of modem that can transfer data at higher speeds.
- **DSL** technology divides the bandwidth of UTP cable into 250 channels. The bandwidth of each channel is 4 kHz. The first channel is used for telephone, and the other channels are used for transmitting information.
- The Types of DSL are: ADSL, SDSL, HDSL, and VDSL.
- ADSL can transfer data **downstream** at rates of 1.5 to 8 Mbps.
- ADSL can transfer data **upstream** at rates of 16 to 640 Kbps.
- A **cable modem** uses a cable TV network to connect residential computers to the Internet.
- The **head end** of a cable TV system uses TV channels to transmit information to a cable modem at the subscriber site.
- Each cable TV channel requires 6 MHz of bandwidth.
- Connecting a computer to a cable modem requires a **10BaseT** network card.

- More than one station can be connected to a cable modem by using a hub or repeater.
- A cable modem operates in layer 1, layer 2, and layer 3 of the OSI model.
- **IEEE 802.14** is the standard for cable modem.
- The **Integrated Services Digital Network (ISDN)** provides end-to-end digital connection.
- **Narrowband ISDN** offers Basic Rate Interface (BRI) and Primary Rate Interface (PRI).
- The **Basic Rate Interface (BRI)** is made of two B-channels and one D-channel. B-channel data rate is 64 Kbps and D-channel is 16 Kbps.
- The **Primary Rate Interface (PRI)** is made up of 23 B-channels and one 64 Kbps D-channel.
- The **BRI** offers the telephone subscriber line two telephone lines and one data line. The two telephone lines can be combined and used as a 128 Kbps data line.
- Applications of the **PRI** are: (1) to connect two central switches together and (2) to be used as a T1 link.
- A **Network Termination Type 1 (NT1)** device is connected to a subscriber telephone line in order to provide service for two telephones and one computer.

Key Terms

10BaseT

Amplitude Shift Keying (ASK)

Asymmetrical Digital Subscriber Line

Basic Rate Interface (BRI)

Baud Rate

Broadband ISDN (B-ISDN)

Cable modem

Coaxial Amplifier

Constellation Diagram

Data Rate

Demodulation

Digital Subscriber Line (DSL)

Downstream

Drop Cable

External Modem

Feeder Cables

Frequency Shift Keying (FSK)

Head End

IEEE 802.14 Standards

Integrated Services Digital Network (ISDN)

Internal Modem

Modem

Modulation

Narrowband ISDN (N-ISDN)

Network Terminator Type 1

Phase Shift Keying (PSK)

Primary Rate Interface (PRI)

Quadrature Amplitude Modulation (QAM)

Trunk Cable

Upstream

Review Questions

• Multiple Choice Questions

1. A modem converts _____.
   - a. analog signals to digital
   - b. digital signals to analog
   - c. a and b
   - d. analog to analog

2. _____ is responsible for developing standards for modems.
   - a. ITU
   - b. IEEE
   - c. EIA
   - d. ISO

3. The maximum theoretical data rate for a modem is _____.
   - a. 33.6 Kbps
   - b. 56 Kbps
   - c. 28 Kbps
   - d. 24 Kbps

4. _____ is the latest technology in modems.
   - a. xDSL
   - b. Cable modem
   - c. Regular modem
   - d. LAN

5. ADSL uses _____ encoding.
   - a. QAM
   - b. DMT
   - c. PSK
   - d. ASK

6. Cable TV is designed to transmit a/an _____ TV signal.
   - a. baseband
   - b. broadband
   - c. analog
   - d. digital

7. What is the data rate of a communication channel with a bandwidth of 40 kHz and each signal represented by 4 bits? _____
   - a. 40 Kbps
   - b. 80 Kbps
   - c. 160 Kbps
   - d. 10 Kbps

8. QAM modulation is a combination of _____.
   - a. ASKand FSK
   - b. ASK and PSK
   - c. PSK and FSK
   - d. none of the above

9. Which of the following devices uses twisted-pair cable? _____
   - a. Cable modem
   - b. DSL modem
   - c. 10Base2
   - d. 10Base5

10. What type of modulation method is used in cable modems for downstream transmission? _____
    - a. DMT
    - b. QAM
    - c. QPSK
    - d. ASK

11. What type of modulation is used in cable modems for upstream transmission? _____
    - a. QAM
    - b. DMT
    - c. QPSK
    - d. FSK

12. DSL operates with a/an _____.
    - a. analog signal
    - b. digital signal
    - c. optical signal
    - d. baseband signal

13. What is the bandwidth of each TV channel? _____
    - a. 4 MHz
    - b. 2 MHz
    - c. 6 MHz
    - d. 1 MHz

14. What is the lowest frequency of TV Channel 1? _____
    - a. 40 MHz
    - b. 50 MHz
    - c. 60 MHz
    - d. 30 MHz

● **Short Answer Questions**

1. What does modem stand for?
2. Explain the function of a modem.
3. Define data rate.
4. Define baud rate.
5. Explain ASK modulation.
6. Explain FSK modulation.
7. Describe PSK modulation.
8. Modem speed is represented by _____.
9. What are the speeds of modems currently being produced?
10. Explain 56 Kbps modem operation.
11. Distinguish between data rate and baud rate.
12. Draw a constellation diagram for 32-QAM using two amplitudes.
13. What does DSL stand for?
14. What does ADSL stand for?

15. Explain ADSL operation.

16. What type of modulation is used for ADSL?

17. Explain xDSL?

18. Why can ADSL transfer information faster than a modem?

19. Is ADSL dependent on cable length?

20. What type of cable is used for ADSL?

21. What are the components of a cable TV system?

22. What does HFC stand for?

23. What is the bandwidth of a TV channel?

24. What type of modulation is used in a cable TV modem for upstream transmission?

25. What is the type of modulation is used in cable TV for downstream transmission?

26. What type of NIC is used in a computer connected to a cable TV modem?

27. List the devices that can be connected to a cable modem.

28. What is the baud rate of ASK with a data rate of 600 bits per second?

29. What is the data rate of a modem using frequency shift keying with a baud rate of 300 signals per second?

30. What is the data rate of a QAM signal with a baud rate of 1200 and each signal represented by four bits?

31. Calculate the number of bits represented by each signal for a modem using PSK modulation with a data rate of 2400 bps and a baud rate of 600.

32. How many bits per signal can be represented by a 32-QAM signal?

33. Calculate the baud rate of a 32-QAM signal with a data rate of 25 Kbps.

34. What does ISDN stand for?

35. List the types of ISDN.

36. How many channels does Basic Rate Interface have?

37. List the data rate of the B-channel and the D-channel for BRI.

38. What is the data rate of the D-channel for PRI?

39. How many devices can be connected to BRI ISDN?

chapter 8

Ethernet and IEEE 802.3 Networking Technology

OBJECTIVES

After completing this chapter, you should be able to:

- Distinguish between Ethernet and IEEE 802.3
- Describe Ethernet access method
- Discuss the function of each field in Ethernet frame format
- Distinguish between Unicast address, Multicast address, and Broadcast address
- Explain the different types of Ethernet media

INTRODUCTION

Ethernet was invented by the Xerox Corporation in 1972. It was further redefined by Digital, Intel, and Xerox in 1980 and renamed Ethernet version I or DIX (Digital, Intel and Xerox). The IEEE (Institute of Electrical and Electronic Engineers) was assigned to develop a standard for Local Area Networks. The committee that standardized Ethernet, token ring, fiber-optic and other LAN technology called it "802." Ethernet can be the least expensive LAN to implement. The IEEE developed the standards for Ethernet in 1984. It is called IEEE 802.3 and uses the bus topology, shown in Figure 8.1. DIX revised version I to be compatible with IEEE standards and named it Ethernet version II.

Figure 8.2 shows how Ethernet fits into the OSI model. The data link layer is divided into two sublayers, the **Logical Link Control (LLC)** and the **Media Access Control (MAC)**. The function of the LLC is to establish a logical connection between source and destination. The IEEE standard for LLC is IEEE 802.2. The function of the Media Access Control is to access the network, which uses CSMA/CD (Carrier Sense Multiple Access with Collision Detection).

8.1   Ethernet Operation

Each network card has a physical address. When a station transmits a frame on the bus, all stations connected to the network will copy the frame. Each station

123

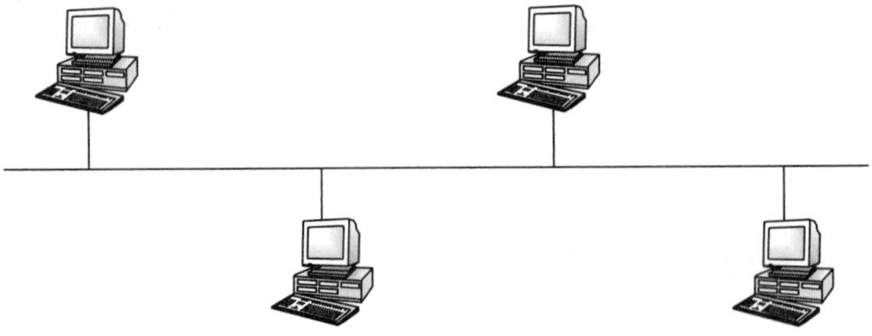

FIGURE 8.1
Ethernet bus
topology

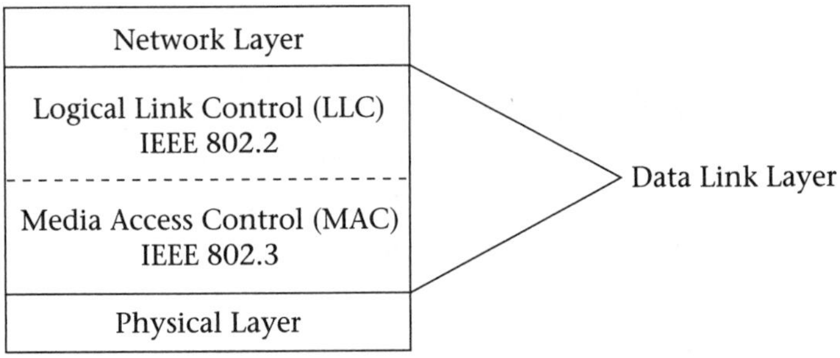

FIGURE 8.2
Ethernet reference
model

checks the address of the frame, and if it matches the station's NIC address, it accepts the frame. Otherwise the station discards the frame.

In an Ethernet network, each station uses CSMA/CD protocol to access the network in order to transmit information. CSMA/CD works as follows:

1. If a station wants to transmit, the station senses the channel (listens to the channel). If there is no carrier, the station transmits and checks for a collision as described in part 2. If the channel is in use, the station keeps listening until the channel becomes idle. When the channel becomes idle, the station starts transmitting again.

2. If two stations transmit frames at the same time on the bus, the frames will collide. The station which detected the collision first sends a jamming code on the bus (a jam signal is 32 bits of all ones), in order to inform the other station that there is a collision on the bus.

3. The two stations, which enveloped in collision, wait according to back-off algorithm (a method used to generate waiting time for stations that were involved in a collision), then start retransmission. Figure 8.3 shows the flowchart of CSMA/CD.

8.2 IEEE 802.3 Frame Format

A block of data transmitted on the network is called a frame. Figure 8.4 shows the IEEE 802.3 frame format.

FIGURE 8.3
CSMA/CD
flowchart

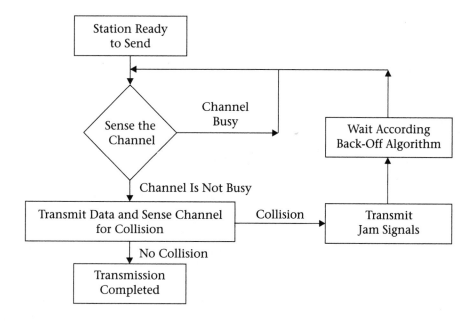

FIGURE 8.4
IEEE 802.3 MAC
and LLC frame
formats

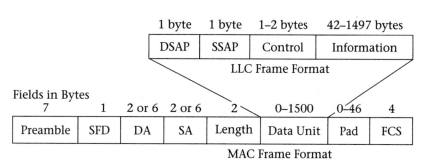

Most network manufacturers use IEEE standards. The items listed below describe each field of IEEE 802.3 frame format.

Preamble: The preamble provides signal synchronization and consists of seven bytes of alternating 1 and 0 bits.

Start of Frame Delimiter (SFD): The SFD represents the start of a frame and is always set to 10101011.

Destination Address (DA): The destination address is the hardware address of a recipient station and it is six bytes (48 bits). This address is a unique address (in the entire world). The hardware address of the Network Interface Card (NIC) is also called a MAC address (media access control) or physical address. The IEEE oversees the physical addresses of NICs world-wide by assigning 22 bits of physical address to the manufacturers of network interface cards. The 46-bit address is burned into the Read Only Memory (ROM) of each NIC and is called the universal administered address. Figure 8.5 shows the format of the destination address. The destination address can have the following types of addresses:

FIGURE 8.5
Format of physical address

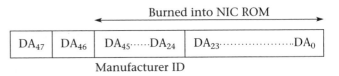

Manufacturer ID

- Recipient is an individual station (**unicast**).
- Recipients are groups of stations (**multicast**).
- Recipients are all stations in the network (**broadcast**). The 48 bits of destination address are all set to ones, meaning that the DA address is FFFFFFFFFFFF Hex for a broadcast address.
- In Figure 8.5 the DA_{47} set to zero means that the address is used for an individual recipient of the frame. A multicast address is a special address which is assigned by the network operating system to a group of stations. When a frame is transmitted to the network with a multicast address, stations having multicast addresses will accept the frame. The multicast address is assigned by IEEE to the manufacturer of NICs. For example, the multicast address for Cabletron NICs is 010010FFFF20 Hex.
- DA_{46} set to one means that the address is a universally administered address.
- DA_{46} set to zero means that the address is locally administered, and a network administrator can assign an address to the NIC. This type of addressing is used only in a closed network.
- DA_{45} to DA_{24} represent the manufacturer identifier.
- DA_{45} to DA_0 are burned into the ROM of the NIC and indicate the physical address of the card.

Source Address (SA): The SA shows the address of the source from which the frame originated.

Length Field: This two-byte field defines the number of bytes in the data field.

Data Field: According to Figure 8.4, the data field contains the actual information. The IEEE specifies that the minimum size of a data field must be 46 bytes, and the maximum size is 1500 bytes.

Pad Field: The IEEE standard defines that the data field must be at least 46 bytes. If information in the data field is less than 46 bytes, extra information is added in the pad field to increase the size to 46 bytes.

Frame Check Sequence (FCS): The FCS is used for error detection to determine if any information was corrupted during transmission. IEEE uses CRC-32 for error detection.

Destination Service Access Point (DSAP): The MAC layer passes information to the LLC layer, which must then determine which protocol the incoming information belongs to, such as IP, NetWare or DecNet.

Source Service Access Point (SSAP): The SSAP determines which protocol is sent to the destination protocol, such as IP or DecNet.

Control Field: The control field determines the type of information in the information field, such as the supervisory frame, the unnumbered frame, and the information frame.

8.3   Ethernet Characteristics

Table 8.1 shows Ethernet characteristics. The gap between each frame should not be less than 9.6 msec. A station can have a maximum of 10 successive collisions. The size of the jam signal is 32 bits of all 1s. The maximum size of the frame is 1512 bytes including the header. Slot time is the propagation delay of the smallest frame. The smallest frame is 512 bits and each bit time is 10^{-7} second, therefore the propagation delay is 512 bit time.

TABLE 8.1   Ethernet Characteristics

Data Rate	10 Mbps
Encoding	Manchester Encoding
Slot Time or Propagation Delay	512 bit time
Interframe Gap	9.6 msec
Backoff Limit	10
Jam Size	32 bits
Maximum Size	1512 bytes
Minimum Frame Size	64 bytes

8.4   Ethernet Cabling and Components

The Ethernet network uses three different media, called 10BaseT, 10Base2 and 10Base5. Figure 8.6 illustrates the ports of an NIC which are used to connect a computer to a network. There are three different types of connectors that come with an NIC. They are BNC, RJ-45 and DIX. RJ-45 is used for 10BaseT connections, DIX is used for 10Base5 connections, and BNC is used for 10Base2 connections, all of which are described in the next section.

FIGURE 8.6
Network Interface
Card (NIC)

RJ-45

DIX

BNC

*BNC British Naval Connector
*RJ-45 Register Jack

ThinNet The specifications of **10Base2 (ThinNet)** as shown in Figure 8.7 are as follows:

- 10Base2 uses thin coaxial cable with BNC connectors.
- The maximum length of one segment is 185 meters.
- The maximum length of a network cable is 925 meters, by using four repeaters.
- The transceiver is built into the NIC.
- The minimum distance between T-connectors is 0.5 meters.
- No more than 30 connections are allowed per segment.
- The first and the last device on each segment must be terminated with a 50-Ω resistor called a BNC terminator. The function of a terminator is to prevent signal reflection on the cable.
- A T-connector must plug directly into the Ethernet device.

ThickNet 10Base5 is occasionally used for network backbones. The transceiver is a separate component attached to a coaxial cable, as shown in Figure 8.8 on page 130. The function of a transceiver is to transmit information on the network, receive information from network, and detect collision.

The specifications of **10Base5 (ThickNet)** are as follows:

- The maximum length of one segment is 500 meters (without repeater).
- Devices are attached to the backbone via a transceiver.
- The maximum length of the Attachment Unit Interface cable (AUI) is 50 meters.
- The minimum distance between transceivers is 2.5 meters.

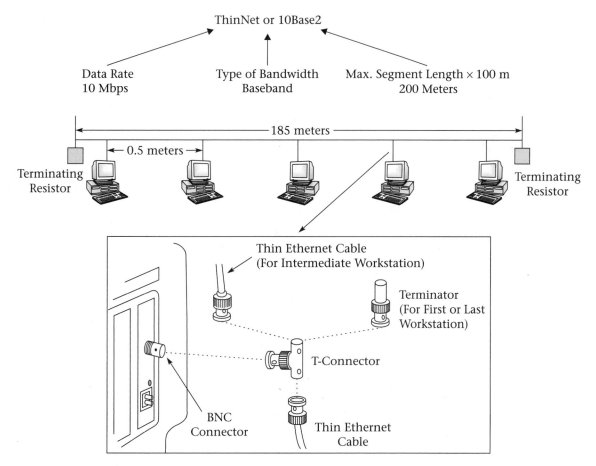

FIGURE 8.7 Components of ThinNet

- No more than 100 transceivers are allowed in one segment of the network.
- Both ends of the segments must be terminated by a 50-Ω resistor.

Thick Coax Transceiver with Signal Quality Error (SQE)   Figure 8.9 shows a thick coax transceiver. Some transceivers come with a single port and others have two or four ports for output. The function of a transceiver is to interface the Ethernet coax cable with an Ethernet Network Interface Card.

The objective of **Signal Quality Error (SQE)** is to inform the station that the collision detection section of the transceiver is working properly. A user selectable switch is provided, permitting the network manager to disable or enable SQE. Figure 8.10 shows the internal architecture of a transceiver. The function of the jabber control is to disconnect the transceiver from the cable in case of a short circuit in the transceiver.

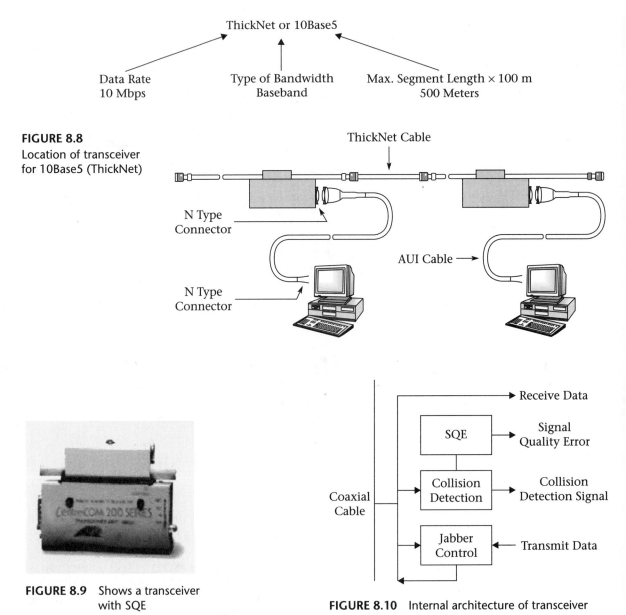

ThickNet or 10Base5

Data Rate
10 Mbps

Type of Bandwidth
Baseband

Max. Segment Length × 100 m
500 Meters

FIGURE 8.8
Location of transceiver
for 10Base5 (ThickNet)

ThickNet Cable

N Type
Connector

AUI Cable →

N Type
Connector

FIGURE 8.9 Shows a transceiver
with SQE

Receive Data

SQE → Signal
Quality Error

Coaxial
Cable

Collision
Detection → Collision
Detection Signal

Jabber
Control ← Transmit Data

FIGURE 8.10 Internal architecture of transceiver

10BaseT 10BaseT uses UTP cable for transmission media, and all stations are connected to a repeater or hub, as shown in Figure 8.11. The function of a repeater (hub) is to accept frames from one port and retransmit the frames to all the other ports. Table 8.2 shows the pin connection of a RJ-45 connector.

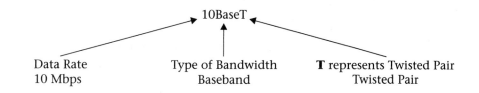

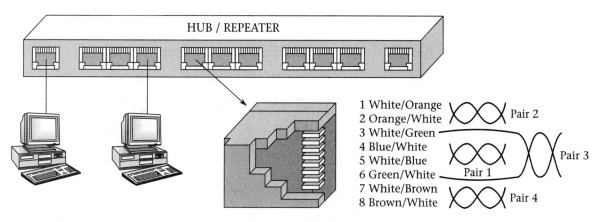

FIGURE 8.11   10BaseT connection

TABLE 8.2   RJ-45 Connector Pins

Pin	Signals
1	RD+
2	RD–
3	TD+
4	NC
5	NC
6	TD–
7	NC

The specifications of **10BaseT** are as follows:

- The maximum length of one segment is 100 meters.
- The transceiver for 10BaseT is built into the NIC.
- The cable used is 22 to 26 AWG unshielded twisted pair cable, category 4 or 5.
- Devices are connected to a 10BaseT hub in a star topology.
- Devices with standard AUI connectors may be attached to the hub by using a 10BaseT transceiver.
- 10BaseT topology allows a maximum of four repeaters connected together, and the maximum diameter is 500 m.

A quick reference chart of IEEE 802.3 networks is provided in Table 8.3.

TABLE 8.3   Comparison of Different Types of IEEE 802.3 Networks

	10Base5 (ThickNet)	10Base2 (ThinNet)	10BaseT Unshielded Twisted Pair
Medium	Coaxial Cable 50 Ω–10 mm (dia)	Coaxial Cable 50 Ω–5mm (dia)	Twisted Pair Category 3, 4, 5
Signal	10 Mbps Baseband/Manchester	10 Mbps Baseband/Manchester	10 Mbps Baseband/Manchester
Maximum Segment	500 m	185 m	100 m
Maximum Distance	2.5 km	0.925 km	500 m
Nodes per Segment	100	30	N/A
Topology	Bus	Bus	Star

Summary

- **Ethernet** or **IEEE 802.3** uses bus topology.
- In the bus topology, the medium is shared by all stations.
- Ethernet uses **Carrier Sense Multiple Access/Collision Detection** (CSMA/CD) to access the network.
- Ethernet data rate is 10 Mbps.
- The maximum size of an Ethernet frame is 1512 bytes.
- An Ethernet network card comes with three types of connectors: **RJ-45**, **BNC**, and **DIX**.
- **ThinNet** or **10Base2** features 10 Mbps, baseband, and 200 meters per segment using thin coaxial cable with BNC connector.

- **10BaseT** is a medium with these features: 10 Mbps, baseband, and using a UTP cable with a RJ-45 connector.
- 10BaseT requires a repeater or hub.
- A **Unicast address** tells you that the recipient of the frame is an individual station.
- A **Multicast address** indicates that the recipient of the frame is a group of stations.
- A **Broadcast address** means the recipient of the frame is every station in the network.
- **ThickNet** or **10Base5** features 10 Mbps, baseband, and 500 meters per segment using thick coaxial cable with a DIX connector.
- 10Base5 is used for the backbone, and an external **transceiver** makes the connection between the NIC and cable.
- There is a maximum of four **repeaters** allowed for use in Ethernet.

Key Terms

Broadcast	Repeater
CSMA/CD	Signal Quality Error (SQE)
Destination Address (DA)	Source Address (SA)
Destination Service Access Point (DSAP)	Source Service Access Point (SSAP)
	10Base2
Ethernet Frame Format	10Base5
Frame Check Sequence (FCS)	10BaseT
IEEE 802.3	ThickNet
LLC Frame Format	ThinNet
Logical Link Control (LLC)	Transceiver
Media Access Control (MAC)	Unicast
Multicast	

Review Questions

- **Multiple Choice Questions**
    1. _____ is the least expensive LAN.
        - a. Ethernet
        - b. Token ring
        - c. a and b
        - d. Gigabit Ethernet

2. The standard for Ethernet is _____.
   a. IEEE 802.3
   b. IEEE 802.4
   c. IEEE 802.5
   d. IEEE 802.2

3. Ethernet uses _____ to access a channel.
   a. CSMA/CD
   b. token passing
   c. demand priority
   d. full-duplex

4. A destination address has _____ bytes.
   a. 2
   b. 3
   c. 6
   d. 8

5. Ethernet uses _____ encoding.
   a. Manchester
   b. Differential Manchester
   c. NRZ
   d. NRZ-I

6. Ethernet networks come with _____ different media.
   a. 3
   b. 6
   c. 8
   d. 2

7. An RJ-45 connector is used for _____.
   a. 10BaseT
   b. 10Base2
   c. 10Base5
   d. none of the above

8. 10Base2 uses _____ cable.
   a. UTP
   b. STP
   c. thin coaxial
   d. thick coaxial

9. The maximum length of one segment of 10Base2 is _____ meters.
   a. 10
   b. 185
   c. 100
   d. 300

10. Which of the following networks requires separate transceivers? _____
    a. 10BBaseT
    b. 10Base5
    c. 10Base2
    d. None of the above

11. What type of Ethernet would you chose for connecting two computers? _____
    a. 10BaseT
    b. 10Base2
    c. 10Base5
    d. None of the above

- **Short Answer Questions**

1. Define the following terms:
   a. 10BaseT   c. 10Base2
   b. 10Base5

2. What is coaxial cable?

3. What do UTP and STP stand for?

4. What is 10BaseT topology?

5. What is a network segment?

6. What does AUI stand for?

7. Explain function of repeater or hub.

8. Show the IEEE 802.3 frame formats and the function of each field.

9. Describe access method for Ethernet.

10. What does CSMA/CD stand for?

11. What is IEEE 802.2?

12. Determine the location of transceiver for the following network interface cards:
    a. 10Base5   c. 10BaseT
    b. 10Base2

13. What is the MAC address?

14. Explain the function of jabber control in a transceiver.

15. Explain collision in Ethernet.

16. What is a jam signal?

17. Explain broadcast address.

18. Describe unicast address.

19. What is the size of a network interface card address?

20. What is application of CRC (Cyclic Redundancy Check)?

21. What is the function of a transceiver?

22. What is SQE used for in a transceiver?

23. Determine the maximum size of a network using three repeaters for the following media:
    a. 10BaseT   c. 10Base2
    b. 10Base5

24. What is the maximum number of repeaters allowed for the following media:

   a. 10Base2   c. 10BaseT

   b. 10Base5

25. What is the maximum size frame for IEEE 802.3?

26. How many bits of a network address represent the manufacturer ID?

27. How do computers distinguish one another on an Ethernet network?

28. What happens when two or more computers simultaneously transmit frames on an Ethernet network?

29. What is the function of the FCS field in the Ethernet frame format?

30. What is the function of the back-off algorithm in an Ethernet network?

31. What is the function of length field in an Ethernet frame?

32. What is the function of a terminating resistor in 10Base5 and 10Base2?

33. List the IEEE sublayers of data link layer.

34. What is function of Pad field in IEEE 802.3 frame format?

chapter 9

Token Ring and Token Bus Networking Technology

OBJECTIVES

After completing this chapter, you should be able to:

- Discuss token ring technology
- Discuss token ring access method
- Explain the frame format of token and the function of each field
- List the components of token ring topology
- Explain token ring MAU and cabling
- Understand the function of active monitor
- Discuss token bus operation

INTRODUCTION

Token ring technology was originally developed by IBM in the 1970s. **Token ring** is a powerful LAN topology that is designed to handle heavy loads. Token ring networking was introduced in 1985, at a data rate of 4 Mbps. The IEEE 802.5 specification was modeled after the IBM token ring. IBM introduced a second type of token ring in 1989 with a data rate of 16 Mbps. Currently, the term token ring is generally used to refer to both IBM's token ring network and IEEE 802.5 networks, as there is little difference between the two types. Token ring technology consists of a ring station and a transmission medium, as shown in Figure 9.1. A ring is a combination function (physically a star and logically a ring topology) that allows a device to connect to the ring. A token ring network uses a wiring concentrator device called a Multistation Access Unit (MAU). The token ring network can be configured with one or more rings and with up to three servers connected to each ring. This will be discussed in more depth in the next section of this chapter.

FIGURE 9.1
Token ring
topology

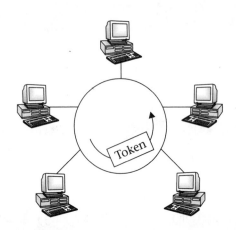

9.1 Token Ring Operation

Figure 9.2 shows the flow of information within token ring networks. Traffic passes through each station on the ring, and each station repeats the information to the next station on the ring.

FIGURE 9.2
Flow of information
on a token ring
network

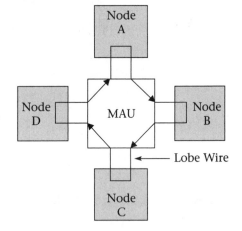

Physically, token ring is a star topology, and electrically it is a ring topology. The **token** itself is a three-byte frame circulating around the network. Any station that wants to transmit information must seize the token, and only then can it transmit information. When a station does not have any information to transmit, it passes the token to the next station. If a station possesses the token and has information to transmit, it inserts information into the token and transmits the frame on the ring. The next station checks the destination address of the frame, and if it matches the station address, it performs following the functions:

● The node copies the frame into its buffer.

- The buffer sets the last two bits of the frame to inform the source that the frame was copied by the destination.
- The frame is retransmitted on the ring.
- The frame circulates on the ring until it reaches the source, which removes the frame from the ring.
- The source releases the token by changing the **Token Bit** (T bit) to one.

9.2 Physical Connections

A ring station is called a **Multistation Access Unit**, or **Multiple Access Unit (MAU)**, as shown in Figures 9.3(a) and 9.3(b). Figure 9.3(a) shows MAU using IBM type connectors, and Figure 9.3(b) shows MAU using RJ-45 connectors. Each computer is connected to the MAU and each MAU can accommodate up to eight stations. **Shielded twisted-pair cabling (STP)** is used for a 16 Mbps token ring network, and **unshielded twisted-pair cabling (UTP)** is used for a 4 Mbps token ring network. When a station is attached to the ring, it performs the following functions:

- The station receives a frame from the ring and passes it to the next station.
- The station then removes its data from the ring.
- Each station repeats the data on the ring, and when the data returns to the originating station, it checks for errors.

FIGURE 9.3(a)
MAU IBM data type connectors

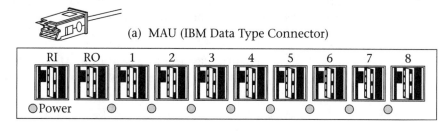

(a) MAU (IBM Data Type Connector)

FIGURE 9.3(b)
MAU RJ-45 connectors

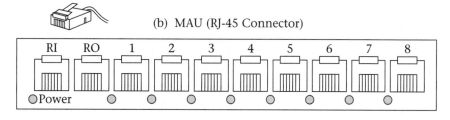

(b) MAU (RJ-45 Connector)

Expanding the Ring A maximum of eight stations can be connected to each MAU. In order to connect sixteen stations to a ring, two MAUs (MAU1 and MAU2) are required. Each MAU has a ring-out port and ring-in port. Connecting ring-out of MAU1 to ring-in MAU2 and then connecting ring-in of MAU1 to ring-out of MAU2 results in a larger ring to which sixteen stations can be connected, as shown in Figure 9.4.

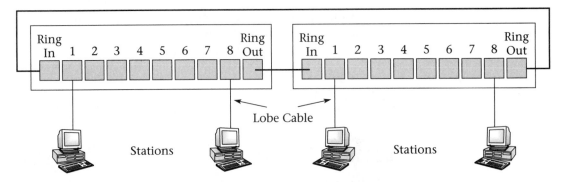

FIGURE 9.4 Connection of two MAUs

9.3 Ring Management

Token ring technology and protocol were designed to make a token ring net-work self-managing. Token ring network management is accomplished by using the token ring network card. All token ring NICs have network management functions, such as adding a new station to the ring without network disruption, generating a token, monitoring activity of the ring, and reporting the loss of a token. These functions are specified by IEEE 802.5.

Error Detection and Correction

One station acts as the **Active Monitor** on a ring. All other stations on the ring are standby monitors. The functions of the active monitor are:

1. To generate a 24-bit token.
2. To broadcast the active monitor MAC frame every seven seconds to all the stations in the ring, which informs stations that there is an active monitor on the ring.
3. To determine the address of any active upstream neighbors.
4. To detect lost tokens and frames.
5. To maintain the ring master and checks for control timing.
6. To purge the ring: the active monitor sends a beacon frame on the ring to inform all stations when there is a problem in the network and token passing has stopped.
7. When an active monitor fails, it is the responsibility of the **Standby Monitor** to become the active monitor.

Adding Stations to the Ring

To add a station to the ring the following process must take place.

1. The station must be physically inserted into the ring.
2. The station then sends multiple MAC frames to the MAU to test the wir-ing lobe.

3. The station must verify it has a unique address by sending a MAC frame with the same source and destination address and check frame status (CFS) field to see if any station copied the frame.

4. The station determines its upstream neighbor and informs its downstream neighbor of its address.

Physically Inserting a Station into the Ring   Figure 9.5(a) shows a ring before insertion of new stations, and Figure 9.5(b) shows stations after insertion to the ring.

FIGURE 9.5(a)
Ring before insertion
of new stations

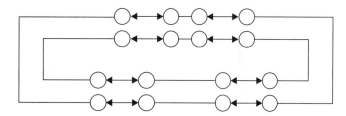

FIGURE 9.5(b)
Ring after insertion
of new stations

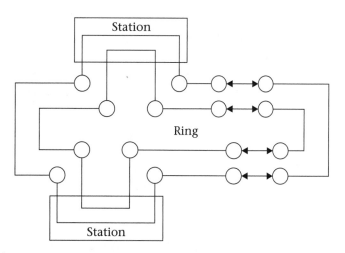

9.4   Token Frame Format

Remember, the token is a three-byte frame, as shown in Figure 9.6, that is passed from station to station on the network. Only one token is allowed on the network at any given time.

The following describe functions of each field of the token frame format:

SD: Start Delimiter of token or frame set to JK0JK000

ED: End of Delimiter and set to JK1JK10E

FIGURE 9.6
Token frame format

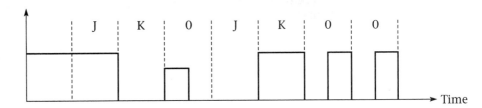

8 bits	8 bits	8 bits
SD	AC	ED

J and K are non-data bits. These bits are Differential Manchester code violations; they do not have mid-point transitions. The signal level for J is the same as for the previous bit and the signal level for K is the opposite of the J signal. This is shown in Figure 9.7. The purpose of this pattern is to keep the SD byte or ED byte from repeating in the information field of token ring frame. The Ending Delimiter field (ED) is JK1JK10E. E is always set to zero, and if any station detects an error, it will set this bit to one to alert the other stations.

FIGURE 9.7
JK0JK00 timing diagram

Access Control Byte (AC)

Figure 9.8 shows the access control byte. The access control field contains three priority bits (it may contain up to eight), and a token bit, which informs the stations whether the data is in token or frame format. The R bits are reserved.

P = Priority bits (000 through 111)

T = Token bit (0 = frame; 1 = token)

R = Reserve bits

FIGURE 9.8
Access control bits

P	P	P	T	M	R	R	R

The **Monitor Bit** (M) is used to prevent frames from circulating onto the ring. When a frame passes through the Active Monitor, the M bit is set to 1. If a frame is passed by an active monitor with the M bit set to 1, the active monitor assumes that the frame has already circulated the ring. The active monitor will then remove the frame from ring.

9.5 IEEE 802.5 Frame Format

A frame is a unit of information that is used by a token ring network for transferring information between stations. The IEEE 802.5 defined the frame format for the token ring network, which is shown in Figure 9.9.

FIGURE 9.9
IEEE 802.5 MAC and
LLC 802.2 frame format

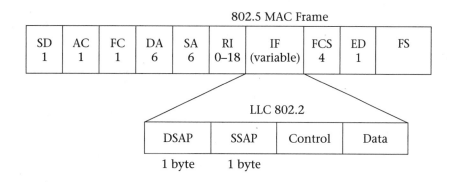

802.5 MAC Frame

SD	AC	FC	DA	SA	RI	IF	FCS	ED	FS
1	1	1	6	6	0–18	(variable)	4	1	

LLC 802.2

DSAP	SSAP	Control	Data
1 byte	1 byte		

The following are the functions of each field of a token ring frame format:

Frame control (FC): FC is 8 bits, as shown in Figure 9.10.

FIGURE 9.10
Frame control (FC)

F F	R R	Z Z Z Z

FF Field: frame type bits

00 = MAC frame
01 = LLC frame
10 = Reserved bit
11 = Reserved bit

RR Reserved Bits: RR is set to 00

ZZZZ Bits: ZZZZ is 0000 (meaning a normal buffer) and ZZZZ is 0001 (meaning an express buffer).

Destination Address Field (DA): The destination address consists of six bytes and is the hardware address of the recipient station. DA uses the same address format as IEEE 802.3.

Source Address (SA): The source address is six bytes and represents the originator of the frame.

Routing Information (RI): Routing information is sometimes included between the source address and the data. This information is optional and not standard in the IEEE 802.5 standard.

Information field (IF): If the information field contains a MAC frame, the frame is called the MAC protocol data unit. If the information field contains an LLC frame, the field is called the LLC protocol unit (LPDU).

Frame Check Sequence (FCS): 4 bytes

Frame Status Field (FS): The frame status is one byte, as shown in Figure 9.11.

FIGURE 9.11
Frame status field

A	C	R	R	A	C	R	R

Access Control (AC) Values:

00 = No station recognized the address and frame

11 = Station recognized the DA and copied the frame

10 = Station recognized the DA but did not copy the frame. This may also indicate that the buffer is full.

01 = Invalid

There are two types of framed data: one is generated by higher layers of protocol and the other is generated by application software. These frames usually contain application data or network commands and are called LLC frames. **MAC frames** contain major vector commands such as active monitor setting, beacon frame, and ring purge frame, and are used for controlling the network.

9.6 Token Ring NIC and Cable Specifications

Figure 9.12 shows a token ring NIC with two types of connectors: the RJ-45 used for UTP cable, and the DB-9 used for STP cable. Table 9.1 shows token ring specifications and Table 9.2 shows token ring cable specifications. Table 9.3 compares the token ring with the Ethernet network.

FIGURE 9.12
Token ring network interface card

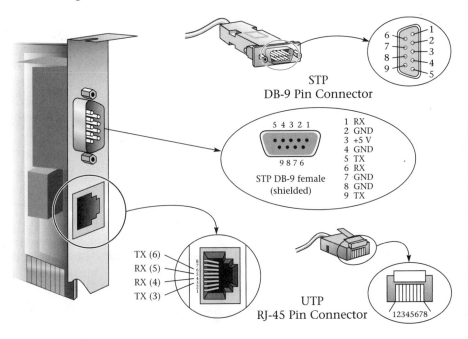

TABLE 9.1 Token Ring Specifications

Maximum Number of Stations	260 stations on one ring using shielded twisted-pair cable (STP), and 72 stations on one ring using unshielded twisted-pair cable (UTP)
Transmission Media	UTP or STP
Data Rate	4 Mbps or 16 Mbps

TABLE 9.2 Token Ring Cable Specifications

Speed/Cable Type	Max. Length of Cable (in meters)	Max. Length Between MAUs (in meters)
4 Mbps/UTP	100	100
16 Mbps/UTP	75	75
4 Mbps/STP	300	200
16 Mbps/STP	100	100

TABLE 9.3 Comparison of Token Ring and Ethernet

	Token Ring	Ethernet
Priority	Yes	No
Routing Information Field	Yes	No
Frame Type	IEEE 802.5	IEEE 802.3
Frame Size	1–18000 bytes	64–1500 bytes
Performance	Deterministic	Variable
Cable	UTP/STP/Fiber/Coax	UTP/STP/Fiber/Coax
Speed	4/16 Mbps	10 Mbps

9.7 Token Bus (IEEE 802.4)

The **token bus** is a type of network that uses bus topology. It uses token passing for accessing the network and the IEEE 802.4 Committee has defined token bus standards.

Figure 9.13 shows four stations connected to a bus. All of the stations are numbered, which indicates a logical ring. The token is passed from station to station,

FIGURE 9.13
Token bus

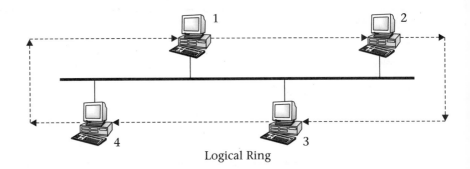

Logical Ring

therefore each station can access the bus by holding the token as it is passed around the logical ring. The applications of token bus network are in manufacturing. Token bus was designed for use in factory automation because of its deterministic delay property, but this technology has not been widely used.

Summary

- The **Token Ring** was initially developed by IBM.
- A token ring uses **Multiple Access Unit (MAU)** as a ring.
- A token ring or **IEEE 802.5** uses token passing as its access method.
- A token is a three-byte frame. The token circulates on the ring and any station that wants to use the ring seizes the token, inserts its frame into the token, and then transmits the frame on to the ring.
- The functions of an **Active Monitor** are to generate the token, detect loss of the token, and control the ring operation by removing frames from the ring and purging the ring.
- The function of **Standby Monitor** is to replace the active monitor should it fail.
- Token ring transmission media are shielded twisted-pair and unshielded twisted-pair cables.
- Token ring topology uses Differential Manchester signal encoding.

Key Terms

Access Control (AC)	Multistation Access Unit (MAU)
Active Monitor	Routing Information (RI)
IEEE 802.5	Standby Monitor
Information Field (IF)	Token Bit
LLC Frame	Token Bus
MAC Frames	Token Frame
Monitor Bit (M)	Token Ring Topology

Review Questions

● **Multiple Choice Questions**

1. Token ring or _____.
   a. IEEE 802.2
   b. IEEE 802.3
   c. IEEE 802.5
   d. IEEE 802.4

2. Token bus or _____.
   a. IEEE 802.5
   b. IEEE 802.4
   c. IEEE 802.3
   d. IEEE 802.2

3. A _____ is a combination function that allows a device to connect to the ring.
   a. bus
   b. ring
   c. token ring
   d. hub

4. A token is a _____-byte frame.
   a. 7
   b. 4
   c. 2
   d. 3

5. A _____ station is called a MAU.
   a. token bus
   b. token ring
   c. bus
   d. ring

6. There is/are _____ active monitor(s) in a ring.
   a. 1
   b. 6
   c. 2
   d. a and c

7. The maximum number of stations which can be connected to ring is _____.
   a. 7
   b. 12
   c. 8
   d. 10

8. Token ring uses _____ encoding.
   a. Manchester
   b. Differential Manchester
   c. a and b
   d. NRZ

9. Which of the following access methods is used in token ring? _____
   a. Token
   b. CSMA/CD
   c. Demand priority
   d. Full-duplex

10. How does a station distinguish between an incoming frame and a token? _____
    a. From the M bit in the AC field
    b. From the T bit of the AC field
    c. From the P bits in the AC field
    d. None of the above

11. Which station generates the token? _____
    a. A station close to the MAU
    b. The station that turns on first
    c. The station that turns on last
    d. All stations in the ring

● **Short Answer Questions**

1. What is the function of a MAU?

2. What is IEEE 802.5?

3. How many stations can be connected to each MAU?

4. What is the access method used by a token ring?

5. How many stations can be connected to a token ring network when:
   a. the stations are connected to MAU by UTP?
   b. the stations are connected to MAU by STP?

6. Show the frame format of the token ring and explain the function of each field.

7. Explain the functions of an active monitor.

8. Describe the function of standby monitor.

9. What is the process for inserting a station into the ring?

10. How can the station distinguish between a token and a frame?

11. What type of signaling is used in a token ring network?

12. Show the IEEE 802.2 frame format and explain the function of each field.

13. How many bits are in a token?

14. What is the difference between a token bus and a token ring?

15. What is purpose of J and K bits in a token ring frame format?

chapter 10

Fast Ethernet Networking Technology

OBJECTIVES   After completing this chapter, you should be able to:

- Discuss Fast Ethernet technology
- Distinguish between the different types of Fast Ethernet media
- Explain the differences and similarities between 100BaseT4, 100BaseTX and 100BaseFX
- Distinguish between different types of repeaters and know the maximum network diameter

INTRODUCTION   Due to recent advances in microprocessor technology, desktop computers have become more powerful, as evidenced in such processors as the Intel Pentium III, the Power PC, and the DEC Alpha. Multimedia applications take advantage of these higher speed desktop computers and offer end users high-end graphics, including 3-D images. The transfer of this type of data between modern computers can become very slow on an Ethernet network. Since the technology of 10 Mbps Ethernet was designed in the mid 1970s, users are now demanding higher data bit rates for the transmission of information.

10.1   Fast Ethernet

Many companies currently use 10BaseT technology, and would like to upgrade their networks to **Fast Ethernet**, which offers a data rate of 100 Mbps. A group of leading network corporations has formed a consortium to draft specifications for Fast Ethernet. This consortium proposed several specifications for Fast Ethernet to the IEEE, as 802.3u. **IEEE 802.3u** is an extension of IEEE 802.3 and was approved as the standard for Fast Ethernet in 1995.

Fast Ethernet is an extension of the Ethernet standard, with a data rate increased to 100 Mbps, using the Ethernet protocol. The goal of Fast Ethernet is to increase the bandwidth of Ethernet networks while using the same CSMA/CD transmission protocol. Using the same protocol for Fast Ethernet allows users to connect an existing 10BaseT LAN to a 100BaseT LAN with switching devices. Figure 10.1 shows the Fast Ethernet protocol architecture.

FIGURE 10.1
Fast Ethernet protocol architecture

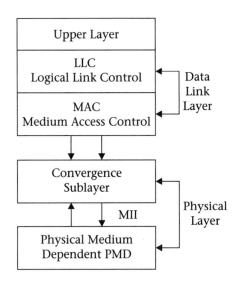

The role of the **Convergence Sublayer (CS)** is to interface the MAC sublayer to the **Physical Medium Dependent (PMD) sublayer** in order to transmit with a higher bit rate, using different media. The **Media Independent Interface (MII)** is defined as the interface between the CS layer and the PMD layer.

10.2 Fast Ethernet Media Types

One of the most popular media for the Fast Ethernet network is unshielded twisted-pair wire, because it is easy to work with and is a less expensive medium. The IEEE has approved specifications for the following types of media for Fast Ethernet:

- 100BaseT4
- 100BaseTX
- 100BaseFX

100BaseT4 100BaseT4 cable is designed to be used with Category 3 unshielded twisted-pair wire (UTP) consisting of four pairs of wires. Categories 4 and 5 can also be used for 100BaseT4. An RJ-45 connector is used to connect a station to the repeater.

100BaseT4 requires fours pairs of twisted-pair wires, as shown in Figure 10.2. Pairs 3 and 4 are used for the transmission of data. One wire from pair 1 and one from pair 2 are used for CSMA/CD. Only three pairs of wires are used for transmission and carrier sense detection. Each pair transmits data at a rate of 33.3 Mbps, which exceeds the 30 Mbps specification for UTP Cat-3. Ternary code (also known as 8B6T) is used to convert eight bits to six bits. This method reduces the data rate of each pair from 33.3 Mbps to 25 Mbps.

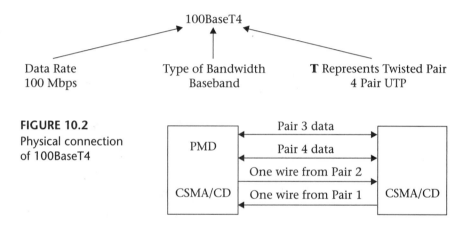

100BaseT4

| Data Rate | Type of Bandwidth | **T** Represents Twisted Pair |
| 100 Mbps | Baseband | 4 Pair UTP |

FIGURE 10.2
Physical connection of 100BaseT4

PMD	Pair 3 data	
	Pair 4 data	
	One wire from Pair 2	
CSMA/CD	One wire from Pair 1	CSMA/CD

100BaseTX 100BaseTX technology supports 100 Mbps over two pairs of Cat-5 UTP or Cat-1 STP wire. Cat-5 UTP cable is the most common media for transmission and is designed to handle frequencies up to 100 MHz. So Manchester encoding, which is used for 10BaseT, is not suitable for 100BaseT, because it doubles the frequency of the original signal. 100BaseT uses 4B/5B encoding with **Multiple Level Transition-3 (MLT-3)** levels for signal encoding. Figure 10.3 shows $(0E)_{16}$ converted to 10 bits 1111011100 using Table 10.1 and then converted to MLT-3. MLT-3 reduces the frequency of the signal by a factor of four.

100BaseFX 100BaseFX technology transfers data at rate of 100 Mbps using *fiber-optic* media for transmission. The standard cable for 100BaseFX is one pair of multimode fiber-optic cables with a 62.5 micron core and 125 micron cladding. The EIA recommends an SC plug-style connector. This SC connector uses push-on and push-off to connect and disconnect from the repeater. Figure 10.4 on page 153 illustrates how a 100BaseFX cable connects to a repeater.

100BaseFX uses 4B/5B encoding with NRZ-I signal encoding. In this type of encoding, four bits of information are converted to five bits, as shown in Table 10.1; the five bits are converted to NRZ-I digital signals, which are then converted to an optical ray for transmission over fiber-optic cable to the receiver.

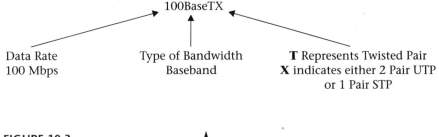

Data Rate | Type of Bandwidth | **T** Represents Twisted Pair
100 Mbps | Baseband | **X** indicates either 2 Pair UTP
| | or 1 Pair STP

FIGURE 10.3
MLT-3 signal for binary value
11110 11100

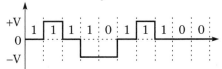

TABLE 10.1   4B/5B Encoding

4 bits Binary	5 bits Symbol	Control Symbols	(5 bits symbol)
0000	11110	Idle	11111
0001	01001	Halt	00100
0010	10100	J	11000
0011	10101	K	10001
0100	01010	T	01101
0101	01011	Set	11001
0110	01110	Reset	00111
0111	01111	Quiet	00000
1000	10010		
1001	10011		
1010	10110		
1011	10111		
1100	11010		
1101	11011		
1110	11100		
1111	11101		

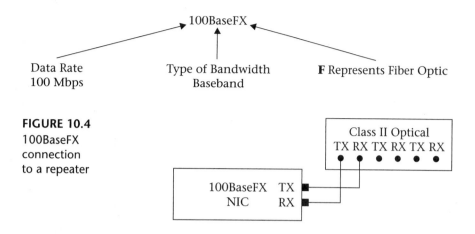

FIGURE 10.4
100BaseFX
connection
to a repeater

At the receiver side, the optical signal is sampled every eight nanoseconds. If there is a change of light (from on to off or from off to on) in the sample, binary one is represented; if there is no change of light, it is represented by binary zero. The conversion from four bits to five bits changes the data rate from 100 Mbps to 125 Mbps. NRZ-I digital encoding reduces the frequency of transmission by half.

10.3 Fast Ethernet Repeaters

Repeaters are used to expand network diameter. There are two types of repeaters used in Fast Ethernet: a Class I repeater and a Class II repeater.

Class I Repeater A **Class I repeater** converts line signals from the incoming port to digital signals. This conversion allows different types of Fast Ethernet technology to be connected to LAN segments. For example, it is possible to connect a 100BaseTX station to a 100BaseFX station by using a Class I repeater, which has a larger internal delay. Only one Class I repeater can be used in a Fast Ethernet segment, as shown in Figure 10.5.

FIGURE 10.5
Class I repeater

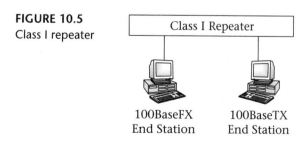

Class II Repeater A **Class II repeater** repeats the incoming signal, sending it to every other port on the repeater. Class II repeaters are used to connect the same media to the collision domain (all stations connected to the repeater are the same type, such as 100BaseTX). Only two Class II repeaters are permitted in one Fast Ethernet segment, as shown in Figure 10.6.

FIGURE 10.6
Application of a
Class II repeater

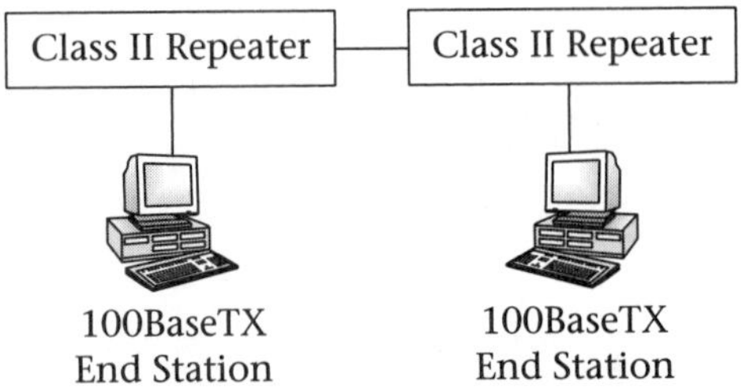

100BaseTX
End Station

100BaseTX
End Station

10.4 Fast Ethernet Network Diameter

Network diameter is a term used to describe the distance between two end stations connected together through a repeater, switch, or bridge. Network diameter is a function of propagation delay. **Propagation delay** is defined as the difference between the transmission time and the receiving time of a signal. This difference is caused by network components such as cable, NIC, and repeaters.

The propagation delay in a network is measured in a unit called **Bit Time**. One bit time is the duration of one bit on the network. For example: Ethernet's bit Time is $1/10^7$ second, and Fast Ethernet's bit time is $1/10^8$ second. In Ethernet networks, the propagation delay is defined as 512 bit times, because the smallest frame is 512 bits and the first bit should reach the destination before the last bit is transmitted by the source.

Table 10.2 shows the propagation delay of Fast Ethernet components.

To calculate network diameter, the total propagation delay of the network components must be less than 512 bit times.

Due to the Electronic Industries Association (EIA) wiring rules for Fast Ethernet, the diameter of Fast Ethernet using twisted-pair wire (100BaseTX and 100BaseT4) is 205 meters. Table 10.3 shows the maximum network diameter using different types of Fast Ethernet repeaters.

The following information can be used to determine the diameter of Fast Ethernet:

- The maximum distance between two repeaters using UTP is five meters.
- The length of all UTP cables should not exceed 100 meters.
- The distance between a repeater and the switch to which it is connected with UTP cable must not exceed 100 meters.
- The maximum distance between two switches connected by fiber-optic cable is 2000 meters (full-duplex operation).

Figure 10.7 shows the connection of 100BaseTX using one Class I repeater.

TABLE 10.2   Propagation Delay of Fast Ethernet Components

Component Type	Bit Times
Two TX NIC or two FX NIC	100
Two T4 NIC	138
One T4 NIC with one TX NIC	127
One TX NIC and one FX NIC	127
100 meters of Cat-3 UTP wires	114
100 meters of Cat-4 UTP wires	114
100 meters of Cat-5 UTP	111
100 meters of STP (IBM type 1)	111
412 meters fiber-optic cable	1 per meter or 412
Class I repeater	140
Class II repeater for TX or FX	92

TABLE 10.3   Maximum Network Diameter

Type of Repeater	100BaseTX or 100BaseT4	100BaseFX
Host-to-host connection (without repeater)	100 meters	412 m
One Class I repeater	200 m	272 m
One Class II repeater	200 m	320 m
Two Class II repeaters	205 m	N/A

FIGURE 10.7
100BaseTX connection
using one Class I repeater

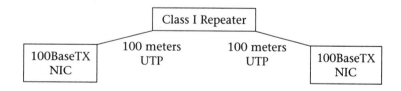

Figure 10.8 shows the connection of 100BaseTX using two Class II repeaters.

FIGURE 10.8
100BaseTX connection
using two Class II
repeaters

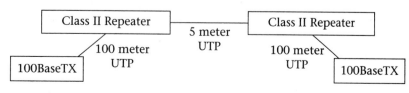

10.5 | Expanding Fast Ethernet

Faster Ethernet allows the use of only two repeaters. The switch can be used to expand Fast Ethernet network diameter. Figure 10.9 shows several segments of Fast Ethernet that are connected together using Fast Ethernet switches.

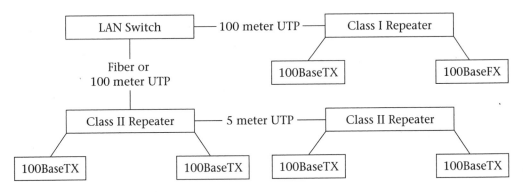

FIGURE 10.9 Expanding Fast Ethernet

Summary

- The data rate of **Fast Ethernet** is 100 Mbps.
- Fast Ethernet can use three types of media: 100BaseT4, 100BaseTX and 100BaseFX.
- Fast Ethernet uses the same frame format as Ethernet.
- 100BaseT4 uses four pairs of **Cat-3 UTP** wires, 100BaseTX uses two pairs of **Cat-5 UTP** wires, and 100BaseFX uses **fiber-optic** cable.
- Fast Ethernet uses **Class I repeaters** to connect NIC cards with different media.
- Fast Ethernet uses **Class II repeaters** to connect stations having the same type of Network Interface Card.
- In Fast Ethernet, only one Class I repeater is allowed.
- Fast Ethernet can use only two Class II repeaters.

Key Terms

Bit Time

Class I Repeater

Class II Repeater

Convergence Sublayer (CS)

Fast Ethernet

Fast Ethernet Access Method

Fast Ethernet Frame Format

IEEE 802.3u

Media Independent Interface (MII)

Multiple-Level Transition-3 (MLT-3)

Network Diameter

100Base FX

100BaseT4

100BaseTX

Physical Medium Dependent (PMD) Layer

Propagation Delay

Review Questions

- **Multiple Choice Questions**
  1. Fast Ethernet or _____.
     - a. IEEE 802.2
     - b. IEEE 802.5
     - c. IEEE 802.3u
     - d. IEEE 802.4

  2. The goal of Fast Ethernet is to increase _____.
     - a. stations
     - b. frequency
     - c. bandwidth
     - d. diameter

  3. The role of _____ is to interface the MAC sublayer to the physical medium dependent layer.
     - a. 100BaseT4
     - b. 100BaseTX
     - c. convergence sublayer
     - d. LLC

  4. _____ is the most popular media for Fast Ethernet.
     - a. UTP
     - b. STP
     - c. Fiber-optics
     - d. Coaxial cable

  5. The data rate of 100BaseTX is _____ Mbps.
     - a. 100
     - b. 10
     - c. 200
     - d. 1000

  6. 100BaseFX uses _____ cable.
     - a. UTP
     - b. STP
     - c. coaxial
     - d. fiber-optic

  7. There are _____ types of repeaters.
     - a. 5
     - b. 2
     - c. 3
     - d. 4

  8. The maximum distance between two repeaters using UTP is _____ meters.
     - a. 10
     - b. 5
     - c. 100
     - d. 200

9. Fast Ethernet's data rate is _____ Mbps.
   a. 100   c. 400
   b. 10   d. 200

10. What type of access method is used in Fast Ethernet? _____
    a. Token   c. Demand priority
    b. CSMA/CD   d. Full-duplex

11. How many Class I repeaters can be used for Fast Ethernet? _____
    a. 1   c. 3
    b. 2   d. 4

12. How many Class II repeaters can be used for Fast Ethernet? _____
    a. 1   c. 3
    b. 2   d. 4

● **Short Answer Questions**

1. Explain the following terms:
   a. 100BaseT4   c. 100BaseFX
   b. 100BaseTX

2. What is the cable type of 100BaseTX?

3. What is the difference between 100BaseTX and 100BaseT4?

4. What is the application of a Class I repeater?

5. What is the application of a Class II repeater?

6. What is the maximum network diameter using two Class II repeaters in a 100BaseT network?

7. Name the IEEE Committee that developed the standard for Fast Ethernet.

8. Explain the access method for Fast Ethernet.

9. What is the function of the convergence sublayer?

10. What are the types of media used for Fast Ethernet?

11. What type of signal encoding is used for 100BaseT4?

12. What type of encoding is used for 100BaseFX?

chapter 11

100 VG–AnyLAN Networking Technology

OBJECTIVES

After completing this chapter, you should be able to:

- Describe 100 VG–AnyLAN node operations
- Describe 100 VG–AnyLAN access methods
- List the types of end stations
- Discuss the operation of 100 VG–AnyLAN nodes
- Understand 100 VG–AnyLAN repeater technology and operation
- Explain the 100 VG–AnyLAN frame format and describe the function of each field

INTRODUCTION

100 VG–AnyLAN means 100 Mbps Voice Grade-Any Local Area Network. 100 VG–AnyLAN is a new technology approved by IEEE as 802.12. Hewlett Packard played a major role in developing and standardizing this new technology. 100 VG–AnyLAN uses star topology, as shown in Figure 11.1, and consists of **end nodes**, **links** and a **repeater (hub)**. The maximum number of nodes which can be connected to each repeater is 32, and the maximum number of levels is five. 100 VG–AnyLAN can be expanded by cascading several repeaters, which is shown in Figure 11.2. The end nodes to each repeater can be workstations with Ethernet NIC or token ring NIC.

An end node can be a PC, a large computer, a user workstation, or a file server. End nodes transmit data to and receive data from a repeater. The end node operates in two modes: **private mode**, which will receive a frame only as its address, and **promiscuous mode**, which will receive all frames.

FIGURE 11.1
100 VG–AnyLAN

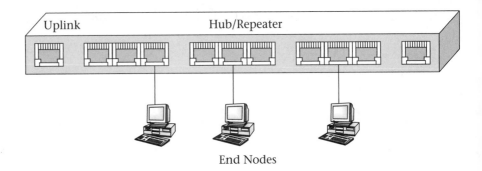

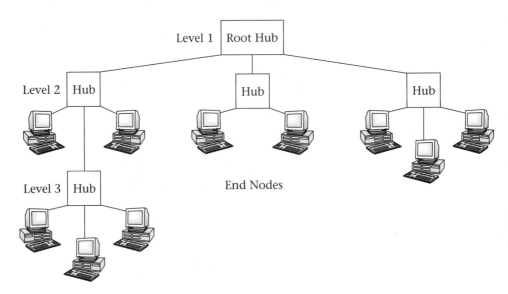

FIGURE 11.2 Large network constructed by cascading several repeaters

11.1 End Node Operation

The end node communicates with the repeater in the following modes:

- **Idle-Up Control:** signals that a node does not have any data to transmit to the repeater
- **Request High Control Signal/Request Low Control Signal:** requests to transmit data
- **Transmitting Data:** end node is transmitting data
- **Listening:** a node, having received an incoming signal from a repeater, goes to listening mode and waits for data

The end node uses Tone 1 and Tone 2 to communicate with a repeater in one of the above listed modes. Tone 1 is a pattern of sixteen ones followed by sixteen zeros. Tone 2 is a pattern of eight ones followed by eight zeros. Table 11.1 shows how end notes make use of the combination of Tone 1 and Tone 2.

TABLE 11.1   Tone 1 and Tone 2 Used for Communication

Tones Transmitted	Receiver is an End Node	Receiver is a Repeater
Tone 1 followed by Tone 1	Idle	Idle
Tone 1 followed by Tone 2	Inform incoming data	Normal priority request
Tone 2 followed by Tone 1	—	High priority request
Tone 2 followed by Tone 2	Link training request	Link training request

11.2   Repeater and Access Method

100 VG–AnyLAN uses **Demand Priority Access Method (DPAM)**, in which nodes make requests for transmission of data to the repeater. The node can request **normal priority** for transmission of normal frames or **high priority** for transmission of sensitive time-dependent frames. The high priority request is granted before a normal request.

A repeater (hub) is an intelligent network controller which uses the "round robin" method to scan each repeater port to manage network access. Each repeater has an **uplink** port that is used to connect to another repeater, as shown in Figure 11.3. Individual nodes on the repeater can be configured to be in private mode or promiscuous mode. Repeaters can also be configured to handle IEEE 802.3 frames or IEEE 802.5 frame formats. The stations connected to each repeater must have the same network cards.

FIGURE 11.3

Connection of two 100 VG–AnyLAN hubs

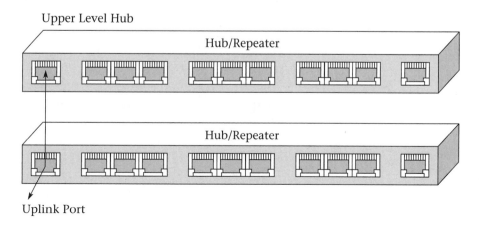

Upper Level Hub

Uplink Port

11.3 100 VG–AnyLAN Transmission Media

The 100 VG–AnyLAN will use four types of cabling: UTP CAT-3, UTP CAT-5, shielded twisted-pair cable, and multimode and single-mode fiber-optics cable. Table 11.2 shows the cable types and the maximum distance of each link.

TABLE 11.2   IEEE 802.12 Cable Specifications for 100 VG–AnyLAN

Cable Types	Maximum Distance
4 pair of UTP CAT-3	100 meters
4 pair of UTP CAT-5	100 meters
2 pair of STP	200 meters
Fiber-optics	2000 meters

11.4 Frame Transmission Method

Data packet transmission uses handshaking sequences; the sending node makes a request and the other side accepts and acknowledges the request. When a node wants to transmit data it performs the following functions:

1. If a node has data to send, it transmits a **Request_Normal** or **Request_High** control signal (normal means normal priority, high means high priority) to the repeater.

2. The repeater scans all ports to find out which node is requesting to transmit data and what the priority of the request is.

3. The repeater selects the end node with high priority request pending (ports are selected in order). The selected port receives acknowledgment from the repeater to start transmitting information.

4. The repeater sends an incoming signal to all other nodes and informs them of the possibility of incoming information. Then the repeater decodes the destination address of the incoming packet.

5. When a node receives an incoming signal, it stops sending requests and listens to the media for a data packet.

6. When the repeater decodes the DA address, it sends a packet to the destination and also sends an idle down signal to other nodes that are not receiving the data packet.

7. When a node receives data, it goes into the **Idle_Down** or **Request_to_Send** mode.

11.5 100 VG–AnyLAN Frame Format

A 100 VG–AnyLAN uses a training frame to set up a link between node and repeater or repeater and repeater, then uses the IEEE 802.3 or IEEE 802.5 frame format for transmitting information between node and repeater. The training frame format is constructed by the MAC layer of nodes. The training frame is transmitted to all repeaters in the network; Figure 11.4 shows the training frame format.

2		6	6	2	2	variable	4	2 bytes
Start of Stream	Preamble	DA	SA	Request Config.	Allow Config.	Data	FCS	End of Stream

FIGURE 11.4 100 VG–AnyLAN training frame fromat

The following bold terms describe the functions of each field of a 100 VG–AnyLAN frame format.

Start of Stream Delimiters (SSD)

High Priority SSD: 0101111100000011
Normal Priority SSD: 0101100000111110

End of Stream Delimiters (ESD)

High Priority ESD: 1111000011000001
Normal Priority ESD: 0000111100111110

Destination Address (DA): The destination address is six bytes.

Null Address: The null address is an address containing all zeros. Null addresses indicate that the frame is not intended for any end node. End nodes cannot be assigned the null address.

Requested Configuration Field Format: The request configuration is two bytes and is used by any nodes or repeaters to inform the higher level repeater about request port configuration. Figure 11.5 shows the request configuration field.

FIGURE 11.5
Request configuration field

R	PP	FF	r r r r r r r r	VVV
1	2	2	8	3 bits

Repeater Bit (R): The repeater bit allows the training initiator to inform the connected higher level repeater of its status (is it a repeater or an end node). An end node may be any MAC-based device such as a user station, bridge, or LAN analyzer.

R = 0: the training initiator is an end node

R = 1: the training initiator is a repeater

Promiscuous Bits(PP): The promiscuous bits provide a means for a lower entity to request the class of unicast addressed packets it wishes to receive.

PP = 00: receive only unicast packets specifically addressed to this end node

PP = 01: reserved for future standardization

PP = 10: receive all packets forwarded by the local repeater (promiscuous mode)

PP = 11: reserved for future standardization

Format bits (FF): The format bits allows the training initiator to request the operational format it wishes to use.

FF = 00: IEE 802.3 format is requested (Ethernet)

FF = 01: reserved for future standardization

FF = 10: IEEE 802.5 format is requested (token ring)

FF = 11: either 802.3 or 802.5 format is acceptable

rrrrrrrr Bits are reserved

VVV Bits are version bits

Allowed Configuration Field Format: Figure 11.6 shows the allowed configuration field format. This is a response to a request configuration field by a higher level entity. The training initiator sets all bits of this field to zero.

FIGURE 11.6
Allowed configuration field format

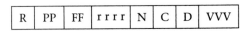

| R | PP | FF | r r r r | N | C | D | VVV |

Access Not Allowed Bit (N): The access not allowed bit is provided for private use to allow the repeater to indicate that the lower entity will not be allowed to join the network for reasons other than configuration (security).

N = 0: access will be allowed only if the configuration is compatible with the network (when C = 0)

N = 1: access will not be allowed

Configuration Bit (C): The configuration bit indicates whether or not the requested configuration is compatible with the network.

C = 0: the configuration is compatible with the network

C = 1: the configuration is not compatible with the network. In this case the FF, PP, and R bits indicate the configuration which would be allowed

Duplicate Address Bit (D): The duplicate address bit indicates whether or not a duplicate address has been detected in the repeater's port address list.

D = 0: no duplicate has been detected

D = 1: a duplicate address has been found

vvv Bits: The vvv bits identify the version of the 802.12 MAC/RMAC training protocol with which the training initiator is compliant. vvv is equal to 001 is the current version.

Training Frame Data Field: This field is set to all zeros.

Frame Check Sequence Field (FCS): IEEE uses a 32-bit cyclic redundancy check (CRC).

Summary

- **100 VG–AnyLAN** means 100 Mbps Voice Grade–Any Local Area Network.
- Hewlett Packard developed 100 VG–AnyLAN and the **IEEE 802.12** Committee approved a standard for 100 VG–AnyLAN.
- 100 VG–AnyLAN consists of End Nodes and a Repeater (Hub); the nodes can be a workstation, a repeater, or a server.
- 100 VG–AnyLAN uses **Demand Priority Access Method (DPAM)**. The repeater uses the "round robin" method to scan each network port for information.
- 100 VG–AnyLAN uses star topology.
- The end node can be a station with an Ethernet NIC or a station with a token ring NIC.
- A repeater can be configured to handle either IEEE 802.3 frame format or IEEE 802.5 frame format.
- End nodes can operate in **Private Mode** (receives frames that have a node address) or **Promiscuous Mode** (receives all frames).
- The maximum number of nodes that can be connected to each repeater is 32.

Key Terms

Demand Priority Access Method (DPAM)

End Nodes

High Priority

Idle_Down

Links

Normal Priority   Request_Normal
100 VG–AnyLAN   Request_to_Send
Private Mode   Tone 1
Promiscuous Mode   Tone 2
Repeater (Hub)   Uplink
Request_High

Review Questions

- **Multiple Choice Questions**

  1. 100 VG–AnyLAN or _____.
     - a. IEEE 802.6
     - b. IEEE 802.14
     - c. IEEE 802.12
     - d. IEEE 802.2

  2. 100 VG–AnyLAN is used in _____.
     - a. WAN only
     - b. MAN only
     - c. LAN only
     - d. none of the above

  3. In a 100 VG–AnyLAN, the end mode operates in _____ ways.
     - a. 2
     - b. 3
     - c. 4
     - d. 1

  4. _____ will receive a frame which has its address.
     - a. Private mode
     - b. Promiscuous mode
     - c. A and b
     - d. None of the above

  5. _____ will receive all frames.
     - a. Private mode
     - b. Promiscuous mode
     - c. A and b
     - d. None of the above

  6. A _____ is an intelligent network controller which uses the "round robin" method.
     - a. server
     - b. client
     - c. UTP
     - d. Repeater

  7. 100 VG–AnyLAN will use _____ types of cabling.
     - a. 2
     - b. 3
     - c. 4
     - d. 5

8. The destination address in 100 VG–AnyLAN frame format contains _____ bytes.
   a. 3
   b. 4

   c. 8
   d. 6

9. The _____ bits identify the version of the IEEE 802.12 MAC/RMAC training protocol.
   a. VVV
   b. duplicate address

   c. format
   d. destination

10. The maximum number of nodes connected to each repeater is _____.
    a. 8
    b. 16

    c. 32
    d. 4

● **Short Answer Questions**

1. What is the topology of the 100 VG–AnyLAN?
2. Explain 100VG–AnyLAN operation.
3. What is the IEEE standard for the 100 VG–AnyLAN?
4. Explain the advantage of the 100 VG–AnyLAN over 100BaseT.
5. List the types of cable for the 100 VG–AnyLAN.
6. Explain the access method for the 100 VG–AnyLAN.
7. Explain promiscuous mode operation for the 100 VG–AnyLAN.
8. Explain private mode operation for the 100 VG–AnyLAN.
9. What is the maximum number of nodes that can be connected to a repeater in the 100 VG–AnyLAN?
10. What is the function of the uplink port of a 100 VG–AnyLAN repeater?
11. Show the training frame format for a 100 VG–AnyLAN and explain the function of each field.

chapter 12

Local Area Network Switching

OBJECTIVES

After completing this chapter, you should be able to:

- Explain switch operation
- Discuss the applications of LAN switching
- Distinguish between symmetric and asymmetric switches
- Understand the technology of a cut-through switch and a store-and-forward switch
- Identify the application of L2 switch, L3 switch, and L4 switch
- Discuss the application of a virtual LAN
- Discuss the application of a firewall

INTRODUCTION

LAN switching is the fastest growing technology in the networking industry. Switches are used to connect LAN segments together in order to increase the network throughput. A **switch** is a device with multiple ports, which accepts packets from one port, examines the destination address, and then transmits the packets to the intended port having a host with the same destination address, as shown in Figure 12.1. Most LAN switches operate at the Data Link layer of the OSI model.

Figure 12.2 shows a symbolic representation of a switch.

12.1 Ethernet LAN Switching

Ethernet is one of the most popular LAN technologies because it uses unshielded twisted-pair cable. However, when the number of stations increases in an Ethernet

FIGURE 12.1
Switch operation

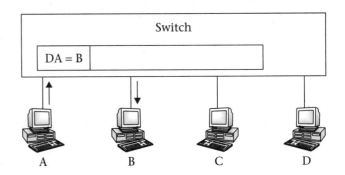

FIGURE 12.2
Symbolic representation
of a switch

LAN, the number of collisions also increases and performance decreases accordingly. In order to increase the performance of an Ethernet LAN, it can be segmented, with the segments connected to switch ports. In Figure 12.3, each segment acts as an independent LAN, and each segment has its collision domain.

A **LAN switch** is similar to multiport bridge. As each LAN frame enters the switch, the switch compares the frame's destination with a table of previously learned addresses, and the frame is sent to the proper port.

If an Ethernet LAN comprises of 20 stations, the bandwidth of each station is equal to the network bandwidth divided by 20. If this LAN were divided into five segments, with each segment connected to a switch, the bandwidth of each station would become the network bandwidth divided by four.

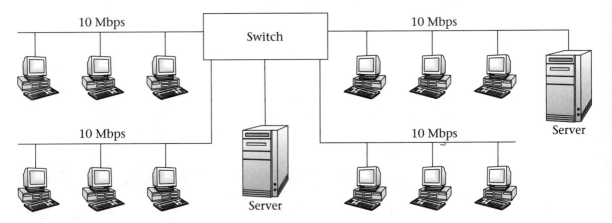

FIGURE 12.3 Connection of LAN segments to a switch

12.2 Switch Classifications

The manufacturers of switches classify the switches based on their applications: symmetrical and asymmetrical.

There are two types of switching:

1. *Symmetric switching* provides switching between segments having the same bandwidth: 10 Mbps to 10 Mbps or 100 Mbps to 100 Mbps, as shown in Figure 12.4.

FIGURE 12.4
Symmetric switch

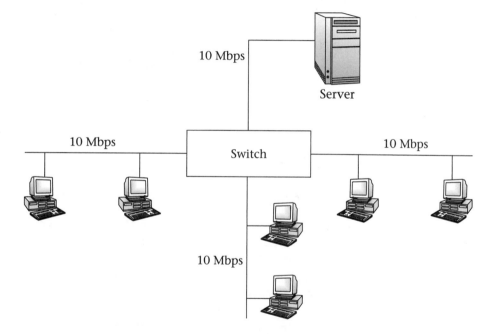

2. *Asymmetric switching* provides switching between segments of different bandwidth: 10 Mbps to 100 Mbps or 100 Mbps to 10 Mbps, as shown in Figure 12.5.

12.3 Switch Operations

A LAN switch uses RISC (Reduced Instruction Set Computer) processors and ASIC (Application Specific Integrated Circuit) processors to increase performance. RISC processors are not as fast as ASIC, but they are less expensive. ASIC switches are custom designed for specific operations, and all of their operations are accomplished through hardware. There are two types of switches:

FIGURE 12.5
Asymmetric switch

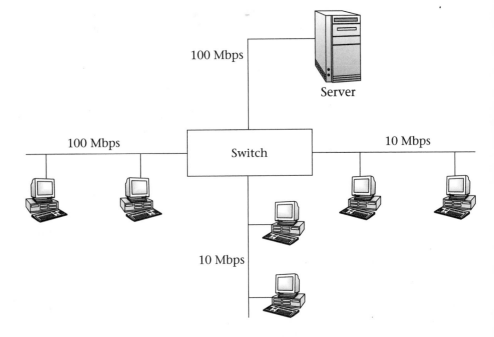

1. *Cut-Through Switch:* A **cut-through switch** reads the first few bytes of the packet to obtain the source and destination addresses. The packets are sent to the destination segment without checking the rest of the packet for errors. The cut-through switch uses an ASIC processor to process the packet.

2. *Store-and-Forward Switch:* The **store-and-forward switch** stores the entire packet, then checks for errors in the packet. If a packet contains errors it is discarded, otherwise the switch forwards the packet to the specified destination. The store-and-forward switch is more suitable for an Ethernet LAN, because it will filter out any corrupted packets to the other segments and, therefore, reduce collision.

12.4   Switch Architecture

Switch architecture is based on the OSI model. The different types of switch architecture are discussed in the following paragraphs.

Layer 2 Switch   A **Layer 2 (L2) switch** operates in the Data Link layer of the OSI model. It is used for network segmentation and for creating workgroups. A Layer 2 switch is similar to a multiport bridge. The frame enters from one port of the switch and is forwarded, based on the MAC address of the frame, to the proper port. A frame with a broadcast address will be repeated to all ports of the switch. A Layer 2 switch learns the MAC addresses of the hosts connected to every port and creates a switching table, which contains the MAC addresses belonging to each port. The switch uses this table to forward the frame to the proper port.

Layer 3 Switch A **Layer 3 (L3) switch** is a type of router that uses hardware rather than software. An L3 switch, sometimes called a routing switch, uses ASIC switching technology such as crossbar switch. This switch operates on the Network layer of the OSI model. The function of an L3 switch is to route the packet based on the logical address (or Layer 3) information. The L3 switch accepts the packet from the incoming port and forwards the packet to the proper port based on a logical address such as an IP address. In order to increase performance, the switch finds the route for the first packet and establishes a connection between the incoming and outgoing ports for transferring the rest of the packets. This is called "route once and switch many." Figure 12.6 shows an application of L3 switches in which they are used to connect the networks of two buildings.

FIGURE 12.6
Connecting the networks
of two buildings using L3
switches

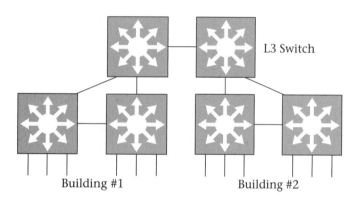

L3 Switch

Building #1   Building #2

Layer 4 Switch A **Layer 4 (L4) switch** operates on the Transport layer of the OSI model. The Internet uses the transport layer with TCP and UDP. TCP is used for reliable communication and UDP is used for unreliable communication. Application protocols running on top of TCP are Telnet, FTP, HTTP, and SMTP. The TCP header contains fields called source port number and destination port number. The source port number identifies the source protocol of an incoming packet and the destination port number identifies the destination protocol for an incoming packet. A Layer 4 switch operates on the port number to forward a packet to the destination. An L4 switch is used for network security and for filtering packets based on application protocol.

12.5   Virtual LAN

A **virtual LAN** (or IEEE 802.10) is a configuration option on a LAN switch that allows network managers the flexibility to group or segment ports on an individual switch into logically defined LANs. There are two immediate benefits from VLAN. First, it provides a way for network administrators to decrease the size of a broadcast domain, for although switches already create many collision domains, they still must pass broadcast frames to every port on the a switch. By creating VLANs and virtual broadcast domains, one can limit the extent to which broadcasts bog down a network. Second, VLANs can provide security options for administrators. A VLAN is one way to prevent hosts on virtual segments from reaching one another. Another application of VLAN is for logical segmentation of workgroups within an organization.

Table 12.1 shows the default connectivity matrix for an **eight-port Ethernet switch**. A + represents connectivity between ports and a – represents no connectivity between ports. This matrix demonstrates that every port has the ability to see and pass packets to every other port on the switch.

TABLE 12.1   Matrix Connectivity of an Eight-Port Switch

Port #	1	2	3	4	5	6	7	8
1	–	+	+	+	+	+	+	+
2	+	–	+	+	+	+	+	+
3	+	+	–	+	+	+	+	+
4	+	+	+	–	+	+	+	+
5	+	+	+	+	–	+	+	+
6	+	+	+	+	+	–	+	+
7	+	+	+	+	+	+	–	+
8	+	+	+	+	+	+	+	–

One advantage of Ethernet switches is that they may be configured to isolate ports from one another, thereby creating virtual LANs. These VLANS can provide isolation from errant broadcasts as well as introduce additional security on the switch.

A network manager can create several VLANs in an Ethernet switch. Table 12.2 shows the matrix connectivity of a switch containing several VLANS, and Table 12.3 shows the VLANs generated by Table 12.2.

12.6   Firewall

Most of an organization's networks are connected together through the Internet or frame relay or leased lines. Security is important and, indeed, essential to

TABLE 12.2   VLAN Matrix Connectivity of an Eight-Port Switch

Port #	1	2	3	4	5	6	7	8
1	+	+	–	–	+	–	–	–
2	+	–	+	–	–	–	–	–
3	+	–	–	+	–	–	–	–
4	+	–	+	–	+	–	+	–
5	–	–	–	+	–	+	–	–
6	+	–	–	–	–	–	+	+
7	+	–	+	–	–	–	–	+
8	+	–	–	–	–	+	+	–

TABLE 12.3   VLANs Generated for Table 12.2

VLAN #	Ports logically connected together
1	1, 2, 5
2	1, 3
3	1, 4
4	1, 3, 5, 7
5	4, 6
6	1, 7, 8
7	1, 3, 8
8	1, 6, 7

prevent hackers from accessing the private network. One of the most popular systems used to implement greater security is the firewall.

A **firewall** is a system (a combination of hardware and software) used to prevent unauthorized access to a private network. All incoming and outgoing packets from the network must pass through the firewall. The firewall examines each packet according to the security criteria and blocks those packets which do not meet the criteria. Following are some firewall techniques:

- **Packet Filtering:** The firewall looks at each packet leaving or entering the network, then accepts or rejects each packet based on the packet header, such as the IP address.

- **Circuit Level Gateway:** When a connection to the network is requested by transmission control protocol or user datagram protocol, the firewall can either give permission for the connection or deny the connection (a TCP connection).

- **Application Gateway:** This applies to applications such as FTP or Telnet server. The firewall can block user access to specific applications on the server.

- **Proxy Server:** Many companies with direct Internet connections use a special server called a proxy server. A proxy server acts as a gateway between an organization network and the Internet. It filters requests that come from the Internet and can block an insider from accessing certain resources from the Internet.

Example: Suppose a corporation requires data connectivity to many business partners over dedicated lines. These circuits could all terminate on an switch. A single firewall can be used to protect the organization's network, as shown in Figure 12.7.

FIGURE 12.7
Application of firewall
and switch

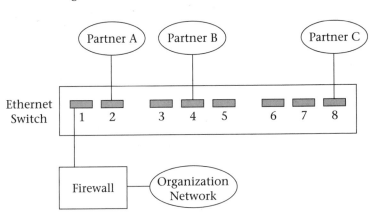

While this configuration would provide protection for the corporation's internal network, it would do nothing to prevent one business partner from trying to connect or hack into another business partner's network. Although it is the responsibility of each business partner to protect its own network, there may be some legal liability to the corporation should an attack or disruption occur because of the interpartner connectivity provided by this switch.

A solution to this problem is neatly provided by the inherent configurability of switches. Network administrators need only to configure multiple virtual LANs to prevent connectivity among the business partner networks. Table 12.4 shows the connectivity matrix of such a virtual LAN. This configuration was adopted to allow a firewall connected to port 1 to see and communicate with the other networks connected to ports 2 through 8. However, ports 2 through 8 cannot see or communicate with one another. The only port they can communicate

TABLE 12.4 Connectivity of VLAN

Port #	1	2	3	4	5	6	7	8
1	–	+	+	+	+	+	+	+
2	+	–	–	–	–	–	–	–
3	+	–	–	–	–	–	–	–
4	+	–	–	–	–	–	–	–
5	+	–	–	–	–	–	–	–
6	+	–	–	–	–	–	–	–
7	+	–	–	–	–	–	–	–
8	+	–	–	–	–	–	–	–

with is port 1. A direct connection to the corporation's internal network is provided by a second Ethernet interface on the firewall, thereby providing connectivity between the corporate network and its business partners.

Summary

- A **switch** accepts a packet from one port and examines the destination address; it then retransmits the packet to the port having a host with the same destination address.
- When the number of users is increased in an Ethernet LAN, the number of collisions will increase. To overcome this problem Ethernet LAN can be segmented, with each segment connected to a port on a switch.
- **Symmetric Switch:** provides switching between LAN segments with the same data rate.
- **Asymmetric Switch:** provides switching between LAN segments with different data rates.
- **Cut-Through Switch:** reads the first few bytes of the frame to determine by which output port the frame must leave.
- **Store-and-Forward Switch:** stores the entire frame and checks for errors. If the frame is corrupted, then it is discarded, otherwise the frame is sent to the proper port for its destination.
- **Virtual LAN (VLAN):** The IEEE 802.9 Committee approved the standard for VLAN. In VLAN the switch port can be enabled and disabled by a network administrator. The administrator can also connect several ports to make a VLAN.
- **Layer 2 Switch:** A multiport device that operates on layer 2 of the OSI model.
- **Layer 3 Switch:** A type of router that uses integrated switching technology.

- **Layer 4 Switch:** A type of switch that operates on layer 4 (Transport layer) of the OSI model.
- A **Firewall** is a system that is used to prevent unauthorized users from accessing an organization's network.

Key Terms

Asymmetric Switch   Layer 4 (L4) Switch

Cut-Through Switch   Proxy Server

Firewall   Store-and-Forward Switch

LAN Switch   Switch

Layer 2 (L2) Switch   Symmetric Switch

Layer 3 (L3) Switch   Virtual LAN (VLAN)

Review Questions

- **Multiple Choice Questions**

  1. The _____ is used to connect segments of a LAN.
     a. router   c. switch
     b. hub   d. gateway

  2. A switch is a device with _____ port(s).
     a. single   c. multiple
     b. two   d. none of the above

  3. _____ provides switching between different bandwidth segments.
     a. Symmetric switching   c. Store-and-forward switch
     b. Asymmetric switching   d. Cut-through switch

  4. A _____ switch reads only the first few bytes of the packet.
     a. cut-through   c. symmetric
     b. store-and-forward   d. asymmetric

  5. Layer 3 switches or routing switches work on the OSI Physical layer, Data Link layer and _____ layer.
     a. Application   c. Presentation
     b. Session   d. Network

6. A _____ is a configuration option on a LAN switch
   a. VLAN
   b. firewall
   c. repeater
   d. router

7. A/an _____ server is one of the firewall techniques.
   a. application
   b. communication
   c. file
   d. proxy

8. A _____ is a system that is used to prevent unauthorized users access to an organization's network.
   a. VLAN
   b. firewall
   c. a and b
   d. router

9. What type of switch is used to connect the segments of a LAN? _____
   a. Layer 2 switch
   b. Layer 3 switch
   c. Layer 4 switch
   d. None of the above

10. A Layer 2 switch operates at the _____.
    a. Physical layer
    b. Data Link layer
    c. Network layer
    d. Application layer

11. What type of switch is used to connect a token ring LAN and an Ethernet LAN? _____
    a. Layer 2 switch
    b. Layer 3 switch
    c. Layer 4 switch
    d. None of the above

12. Which of the following switches is fastest? _____
    a. Store-and-forward
    b. Cut-through
    c. Layer 3
    d. Layer 4

13. Which of the following switches can check for errors in an incoming frame? _____
    a. Store-and-forward
    b. Cut-through
    c. Layer 3
    d. Layer 2

14. What is the application of a Layer 4 switch? _____
    a. Connecting LAN segments
    b. Connecting two different LAN technologies
    c. Used for security
    d. Used for routing

● **Short Answer Questions**

1. Explain switch operation.

2. What is the application of a symmetric switch?

3. What is an application of an asymmetric switch?

4. Explain the operation of a cut-through switch.

5. Explain the operation of a store-and-forward switch.

6. What is a VLAN?

7. In which layer of the OSI model does an L2 switch operate?

8. What are the applications of an L2 switch?

9. In which layer of the OSI model does an L3 switch operate?

10. What is the difference between a router and an L3 switch?

11. What is the application of an L4 switch?

12. Suppose a company has two working groups, A and B. Group A has four computers and group B has three computers; all are connected to an eight-port Ethernet switch. Both groups need to access a common file server; FS1. There is an in-house requirement that group A computers should not be able to see Group B computers in the network.

    a. Draw a diagram showing an Ethernet switch with seven computers and file server.

    b. Show the VLAN connectivity matrix for the above requirements.

chapter 13

Gigabit Ethernet Networking Technology

OBJECTIVES

After completing this chapter, you should be able to:

- Recognize Gigabit standards and Gigabit Ethernet architecture
- Identify the components of Gigabit Ethernet
- Discuss the different types of Gigabit Physical layers
- List some of the applications of Gigabit Ethernet

INTRODUCTION

With recent advances in the PCI bus and CPU, workstations are getting faster. Today's PCI bus can transfer data at gigabit speed. A 64-bit PCI bus runs at 100 MHz and can transfer data at up to 6.4 gigabits. Gigabit Ethernet transfers data at one gigabit per second, or 100 times faster than standard Ethernet. **Gigabit Ethernet** is a new technology that is compatible with Ethernet and Fast Ethernet, and it is used for the backbone with gigabit switches. By adding **Quality of Service (QoS)**, the Gigabit Ethernet will be the next generation of the LAN and the campus backbone.

Gigabit Ethernet is used for the campus **backbone** by connecting gigabit switches together. The switches operate in store-and-forward or cut-through technology. Currently there is no switch mechanism to give priority to the multimedia application frame. The IEEE 802 Committee is developing a new standard called Class of Service (IEEE 802.1p). This is a new protocol that corresponds to the Network layer of the OSI model. IEEE 802.1q and IEEE 802.1p provide tagging for each frame, indicating the priority or class of service desired for the frame to be transmitted. By adding QoS to the Gigabit Ethernet, the Gigabit Ethernet is able to handle all types of data transmission, even voice and video information.

13.1   Gigabit Ethernet Standards

In 1995, the **IEEE 802.3** Committee formed a study group (IEEE 802.3z Task Force) to research the Gigabit Ethernet. The IEEE 802.3z Task Force is developing standards for Gigabit Ethernet. In 1996 the Gigabit Ethernet Alliance was formed, comprising more than 60 companies, to support the development of Gigabit Ethernet.

13.2   Characteristics of Gigabit Ethernet

Gigabit Ethernet will be used for linking Ethernet switches and Fast Ethernet switches and for interconnecting very high-speed servers. Gigabit Ethernet will enable organizations to upgrade their networks to 1000 Mbps while using the same operating systems and the same application software. Following are the characteristics of Gigabit Ethernet:

- Operates at 1000 Mbps
- Uses IEEE 802.3 frame format
- Uses IEEE 802.3 maximum frame format
- Supports full-duplex and half-duplex operation
- Uses CSMA/CD access method for half-duplex operation and supports one repeater per collision domain
- Uses optical-fiber and copper wire for transmission media
- Supports 200 meter collision domain diameters

Gigabit Ethernet Components There are four hardware components required in order to achieve Gigabit Ethernet. They are listed below:

1. Gigabit Ethernet Interface Card
2. Switches that can handle 100 Mbps Fast Ethernet and 1000 Mbps Ethernet
3. Gigabit Ethernet Repeater
4. Gigabit Switch

13.3   Gigabit Ethernet Protocol Architecture

Figure 13.1 illustrates **IEEE 802.3z** Gigabit protocol architecture. The MAC layer offers two types of connections: full-duplex and half-duplex. Half-duplex uses the CSMA/CD access method.

FIGURE 13.1
The IEEE 802.3z Gigabit
Ethernet protocol
architecture

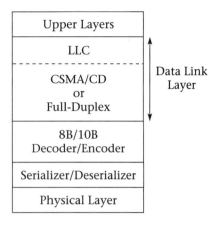

Physical Interface Layer

The Physical Interface Layer defines the physical characteristics of the interface media, including connector type, cable type, transmitter, and receiver. The Gigabit Ethernet Physical layer is designed for the following types of transmission media:

1. **1000BaseLX:** 1000BaseLX is designed for long-wave laser (LW) over Single-Mode Fiber (SMF) and Multimode Fiber (MMF), and having a wavelength of 1300 nanometers.

2. **1000BaseSX:** 1000BaseSX uses short-wave laser (SW) with a wavelength of 850 nanometers over Multimode Fiber.

3. **1000BaseCX:** 1000BaseCX uses twinax cable (150-Ω balanced shielded cable) for transmission media. It supports a 9-pin D-connector.

4. **1000BaseT:** 1000BaseT uses four pairs of CAT-5 UTP cable with a RJ-45 connector for transmission media.

Long-Wave (LW) and Short-Wave (SW) Laser Over Fiber-Optic Cable   Short-wavelength laser is used for short-distance transmission, and long-wavelength laser is used for long-distance transmission. Short-wavelength laser and long-wavelength laser can use multimode fiber. There are two diameters of multimode fiber cables: 50 micron meters, and 62.5 micron meters.

The wavelength of fiber-optic cable is defined by:

$$\text{Wavelength} = \frac{\text{Speed of Light}}{\text{Frequency of Optical Signal}}$$

Modal bandwidth determines the bandwidth of a fiber-optic cable. It also determines the maximum length of the cable, depending on the frequency of the signal in the cable, and is represented by MHz × km. As the data rate increases the length of the fiber-optic cable (km) decreases.

Single-Mode Fiber   Single-mode fiber uses a 9-micron diameter (core) with a 1300-nanometer wavelength laser and is used for long-distance transmission.

1000BaseLX uses a 1300-nanometer long-wavelength laser with SMF and can transmit data to a maximum distance of 5 kilometers.

Multimode Fiber   1000BaseSX uses 850-nanometer short-wavelength laser with MMF and can transmit information a maximum distance of 500 meters.

1000BaseLX uses a 1300-nanometer long-wavelength laser with MMF and can transmit information a maximum distance of 500 meters.

Copper Wire   1000BaseT uses four pairs of unshielded twisted-pair wire with a maximum length of 100 meters.

1000BaseCX uses shield balanced twisted-pair cable for a short distance of 25 meters.

Serializer and Deserializer   The Serializer and Deserializer sublayer accepts information in parallel form from the upper layer (8B/10B decoder/encoder sublayer) and converts it into serial form. It then passes the information to the Physical layer. The Physical layer transfers information to the Serializer/Deserializer in serial form where it is converted into parallel form and passed to the 8B/10B decoder/encoder sublayer.

8B/10B Encoding   8B/10B encoding is used for fiber-optic transmission media. It converts 8-bit information to 10-bit code and then transmits it.

MAC Layer   Gigabit Ethernet supports half-duplex and full-duplex transmission. Gigabit Ethernet half-duplex uses the CSMA/CD access method and full-duplex uses the IEEE 802.3x specification which is a point-to-point connection. CSMA/CD defines the smallest frame for Ethernet as 64 bytes, because the receiver should receive the first bit of the frame before the transmitter completes the transmission. By increasing the speed of the transmission from 100 Mbps to 1000 Mbps, the transmitter can complete transmission before the receiver receives the first bit of the frame. Fast Ethernet overcomes this problem by reducing the size of the cable, and Gigabit Ethernet increases the minimum size of the frame from 64 bytes to 512 bytes by adding carrier extensions to the Ethernet frame. Figure 13.2 shows the minimum frame size for Gigabit Ethernet.

FIGURE 13.2
Minimum frame size format for Gigabit Ethernet

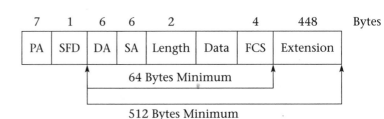

Full-Duplex Transmission   In full-duplex transmission, signals travel in both directions simultaneously. The data rate of the Ethernet is doubled in a full-duplex connection. Full-duplex transmission is used only in point-to-point connections and the CSMA/CD access method is not needed. Figure 13.3 shows a point-to-point connection using full-duplex.

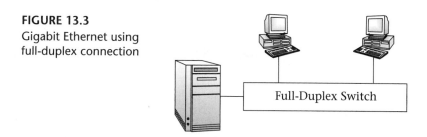

FIGURE 13.3
Gigabit Ethernet using full-duplex connection

13.4   MAC and Physical Layer Architecture

Figure 13.4 shows detailed architecture of the Gigabit Ethernet MAC layer and Physical layer. Information is transferred between the MAC layer, Decoder/Encoder, and Serializer/Deserializer in the form of bytes.

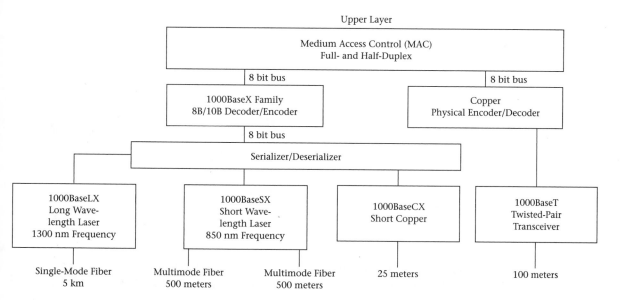

FIGURE 13.4   Components of Ethernet physical layer

Gigabit Ethernet Network Diameters

The Gigabit Ethernet does not support the use of repeaters such as Fast Ethernet and Ethernet. The Gigabit Ethernet supports point-to-point connections using fiber-optic cable, twinax cable and UTP. In Gigabit Ethernet, network diameter can be expanded by using only the Gigabit Ethernet switch. Table 13.1 shows Gigabit Ethernet cable types and distances.

TABLE 13.1 Gigabit Ethernet Cable Types

Standard	Cable Type	Diameter in Microns	Modal Bandwidth MHz × km	Distance in Meters
1000BaseSX	MMF	62.5	160	220
	MMF	62.5	200	270
	MMF	50	400	500
	MMF	50	500	550
1000BaseLX	MMF	62.5	500	550
	MMF	50	400	550
	MMF	50	500	550
	SMF	9	N/A	5000
1000BaseCX	Twinax	N/A		25
1000BaseT	UTP CAT-5			100

MMF means multimode fiber
SMF means a single-mode fiber
SX means short wavelength of 850 nm
LX means long wavelength of 1300 nm
62.5 μm and 50 μm are diameters of the fiber

13.5 Buffered Distributor Device

A **buffered distributor** device is a full-duplex multiport hub. The function of a buffered distributor is to connect several Gigabit Ethernet links together. It operates like a repeater except it stores the Gigabit Ethernet frame and then broadcasts the frame to all ports of the device except the incoming port. The buffer distributor is capable of storing more than one frame in its buffer before forwarding.

13.6 Gigabit Ethernet Applications

Most organizations are currently using Fast Ethernet, which can be upgraded to network with Gigabit Ethernet. The Gigabit Ethernet is used for campuses or buildings that require greater data rate. Gigabit Ethernet applications are:

switch-to-switch, switch-to-server, and repeater-to-switch connections. Gigabit Ethernet is not expected to be used for desktop computers.

Following is a list of Gigabit Ethernet applications:

- Upgrading Fast Ethernet switches to Gigabit Ethernet switches and installing Gigabit Ethernet cards in high-speed servers, as shown in Figures 13.5(a) and 13.5(b). Figure 13.5(a) shows a 100 Mbps network and Figure 13.5(b) illustrates upgrading to 1000 Mbps.

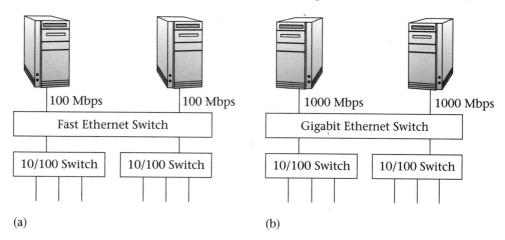

(a)   (b)

FIGURE 13.5(a) and (b)   Upgrading Fast Ethernet to Gigabit Ethernet

- Upgrading a switch to a server link at 1000 Mbps, as shown in Figure 13.6.

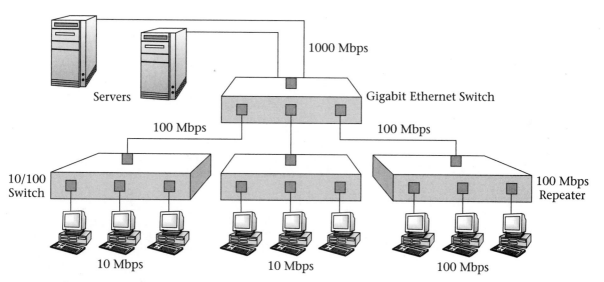

FIGURE 13.6   Upgrading switch to server link to 1000 Mbps

● Upgrading a Fast Ethernet backbone switch from 100 Mbps to 1000 Mbps as shown in Figure 13.7.

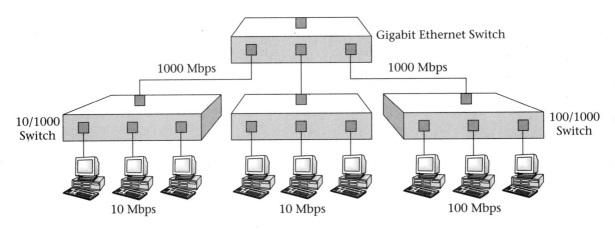

FIGURE 13.7 Upgrading switch Ethernet backbone to 1000 Mbps

● Upgrading the switch to a switch link, as shown in Figure 13.8.

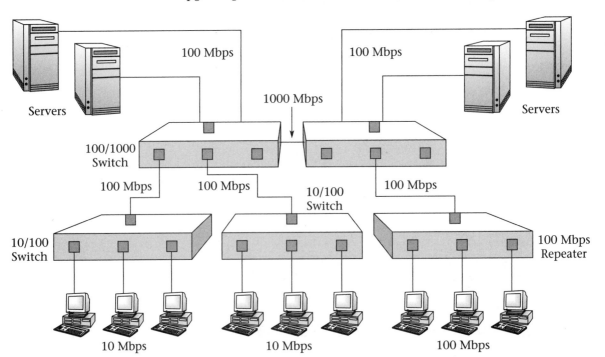

FIGURE 13.8 Upgrading switch to switch link

Summary

- The **IEEE 802.3z** Committee developed the standards for Gigabit Ethernet.
- The **Characteristics of Gigabit Ethernet are:** Data rate of 1000 Mbps; uses IEEE 802.3 format; operates in full-duplex and half-duplex modes; uses optical fiber and copper for transmission media; and uses CSMA/CD for half-duplex operation.
- The Gigabit Ethernet defines the Physical layer for: **1000BaseLX, 1000BaseSX, 1000BaseCX,** and **1000BaseT.**
- **1000BaseLX** means 1000 Mbps, baseband, L for long-wave laser with a wave length of 1300 nm. X means Multimode or Single-Mode Fiber.
- **1000BaseSX** means 1000 Mbps, baseband, S for short-wave laser with a wave length of 850 nm.
- **1000BaseCX** uses shielded twisted-pair cable for transmission media, and **1000BaseT** uses unshielded twisted-pair cable CAT-5.

Key Terms

Backbone	1000BaseCX
Buffered Distributor	1000Base LX
Gigabit Ethernet	1000BaseSX
IEEE 802.3	1000BaseT
IEEE 802.3z	Quality of Service (QoS)
Modal Bandwidth	

Review Questions

- **Multiple Choice Questions**

    1. The data rate of Gigabit Ethernet is _____ Mbps.
       - a. 100
       - b. 1000
       - c. 200
       - d. 10,000

    2. The standard for Gigabit Ethernet is _____.
       - a. IEEE 802.2
       - b. IEEE 802.3
       - c. IEEE 802.3z (Task Force)
       - d. IEEE 802.3u

    3. Gigabit Ethernet uses _____ access method for half-duplex operation.
       - a. CSMA/CD
       - b. token passing
       - c. demand priority
       - d. none of the above

4. Gigabit Ethernet uses _____ encoding.
   a. Manchester
   b. Differential Manchester
   c. 8B/10B
   d. 4B/5B

5. 1000BaseFX uses _____ cable for transmission of data.
   a. UTP
   b. fiber-optic
   c. coaxial
   d. twinaxial

6. What type of protocol should be added to Gigabit Ethernet in order to carry voice and video information? _____
   a. TCP
   b. IP
   c. IEEE 802.1p
   d. RSVP

7. Gigabit Ethernet can operate in _____.
   a. full-duplex
   b. half-duplex
   c. a and b
   d. none of the above

8. Gigabit Ethernet uses the CSMA/CD access method for _____.
   a. half-duplex
   b. full-duplex
   c. a and b
   d. none of the above

9. What is the transmission medium for 1000BaseT? _____
   a. CAT-5 UTP
   b. CAT-4 UTP
   c. Coaxial cable
   d. Fiber-optic cable

10. What is the maximum length of cable used for 1000BaseT? _____
    a. 50 meters
    b. 100 meters
    c. 200 meters
    d. 1000 meters

11. What type of fiber-optic cable is used for Gigabit Ethernet? _____
    a. Multimode fiber
    b. Single-mode fiber
    c. Both a and b
    d. none of the above

12. Gigabit Ethernet is used for _____.
    a. WAN
    b. Campus backbone
    c. MAN
    d. Internet

● **Short Answer Questions**

1. What is the IEEE number for Gigabit Ethernet?

2. What is the data rate for Gigabit Ethernet?

3. What type of frame is used by Gigabit Ethernet?

4. What are the access methods for Gigabit Ethernet?

5. List transmission media for gigabit.

6. Explain the following terms:
   a. 1000BaseCX   c. 1000BaseSX
   b. 1000BaseLX

7. What are the hardware components of Gigabit Ethernet?

chapter 14

LAN Interconnection Devices

OBJECTIVES

After completing this chapter, you should be able to:

- List LAN interconnection devices
- Describe the function and operation of a repeater
- Describe the function and application of a bridge
- Understand the function of a router and the layers of the OSI model that correspond to a router
- Describe the function and application of a gateway

INTRODUCTION

A **LANs interconnection** is the linking of Local Area Networks (LANs) to form a single network. LANs on different floors of a building or LANs in separate buildings can be connected so that all computers in the site are linked.

There are two reasons for linking LANs together: one is to expand the geographic coverage of the network, and the other reason is to divide the traffic load by creating internetworking.

The devices discussed in this chapter are used for linking LANs together and can be distinguished by the OSI layer at which they are operating.

14.1   Repeaters

A **repeater** is a device that is used to connect several segments of a LAN in order to extend the allowable length of the network. A repeater accepts traffic from its input port and then retransmits the traffic at its output port. A hub is a multiple output repeater. A repeater works in the Physical layer of the OSI model. See Figure 14.1, which shows a repeater connecting two segments of a LAN together.

FIGURE 14.1
Two segments of
a LAN connected
by a repeater

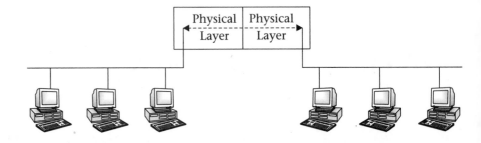

14.2 Bridges

A **bridge** is used to connect different segments of a network together (Homogeneous Network). A bridge operates in the Data Link layer, as shown in Figure 14.2. Bridges forward frames based on destination address of the frame, and can control data flow and detect transmission errors.

FIGURE 14.2
OSI reference
model of a bridge

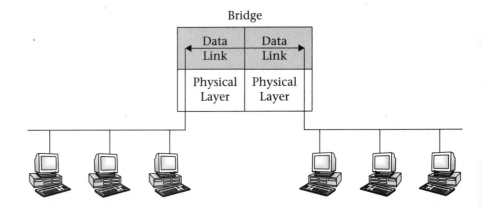

Function of a Bridge The function of a bridge is to analyze the incoming destination address of a frame and make a forwarding decision based on the location of the station. Figure 14.3 shows a bridge that is used to connect two Ethernet LANs together. For example: If station A sends a frame to station B, the bridge gets the frame and sees station B in the same segment as station A and discards the frame. The bridge forwards the data from one LAN to another without alteration of the frame. Bridges allow network administrators to segment their networks transparently. This means that the individual station does not need to know that there is a bridge in the network.

Bridges are capable of filtering. **Filtering** is useful for eliminating unnecessary broadcast frames. Bridges can be programmed not to forward frames from

FIGURE 14.3

Connection of two segments of Ethernet by using a bridge

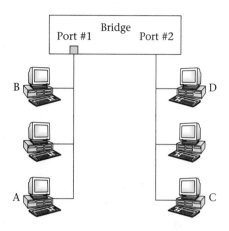

specific sources. By dividing a large network into segments and using a bridge to link the segments together, the throughput of the network will increase. If one segment of the network has failed, other segments connected to the bridge can keep the network alive.

Bridges extend the length of the LAN. While stations A and B are communicating with each other, stations C and D can communicate with each other too.

There are two types of bridges. One type is the **Transparent Bridge** or **Learning Bridge** and the other is the **Source Routing Bridge**.

Learning Bridge or Transparent Bridge: The learning bridge requires no initial programming. It can learn the location of a device by accepting a frame from the network segment and recording the NIC address and the port number. The frame comes to the bridge, which then retransmits the frame to all the segments of the network except the segment which received the frame. By using this method the learning bridge learns which station is connected to which segment of the network.

Source Routing Bridge: The frame contains the entire route to the destination. A source routing bridge is used for a Token Ring Network because a token ring frame has a field that specifies the route of the frame.

14.3 Routers

Routers are more complex internetworking devices than bridges. The **router** works in the Network Layer of the OSI model to route a frame from one LAN to another LAN, as shown in Figure 14.4. To do this, a router must recognize each network layer of the LAN segments connected to the router. Therefore, a router that recognizes multiple network layers is called a multiport protocol router.

FIGURE 14.4
OSI reference model
of a router

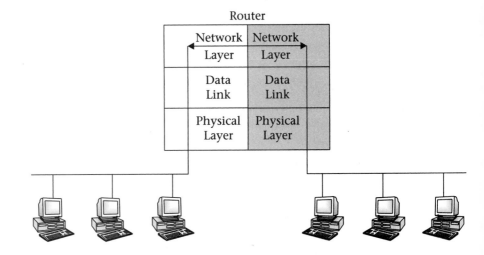

The main function of a router is to determine the optimal data path and transfer the information in that path. Figure 14.5 shows how routers are used to connect several LANs together at different locations. Another function of a router is to convert one type of frame to another type. Station B is connected to a Token Ring Network and has a frame for station A. Router C is capable of converting the token ring frame format to an Ethernet frame format.

FIGURE 14.5
Several LANs connected
together using routers

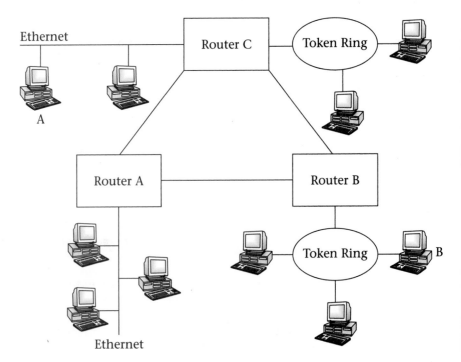

Routers are classified as two types based on their operation. These types are listed below:

1. **Static Router:** In a static router, the routing table is administered manually by the network administrator who determines the route.
2. **Dynamic Router:** The dynamic router sets up its own routing table and updates the routing table automatically. The dynamic router also exchanges information with the next router on the network.

14.4 Gateways

Gateways operate up to the Application layer, as shown in Figure 14.6. The application of a **gateway** is to convert from one protocol to another protocol. In Figure 14.6, a network with IBM SNA architecture is connected, through a gateway, with a LAN running the TCP/IP protocol.

FIGURE 14.6
OSI reference model for a gateway

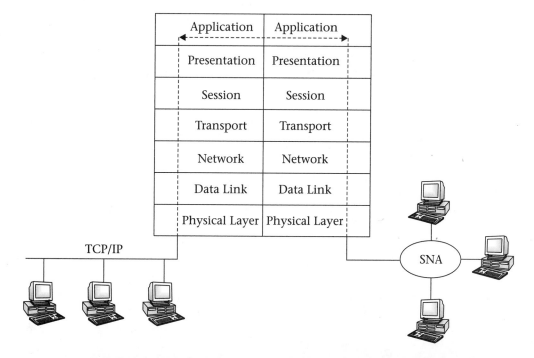

14.5 CSU/DSU

A Channel Service Unit/Data Service Unit (CSU/DSU) is used to connect a LAN to a WAN link. Figure 14.7 shows the application of CSU/DSU. In Figure 14.7, office LANs are connected together by a router, and the router is connected

FIGURE 14.7
Application of a
CSU/DSU

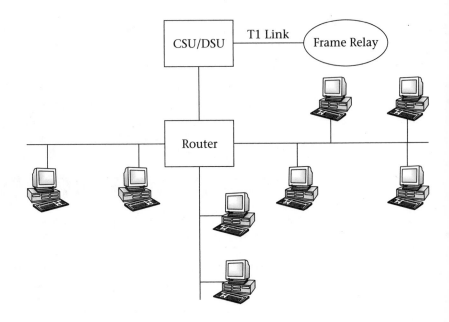

through a T1 link to the frame relay network. The format of the frame and signal types is different from LAN to WAN, therefore, a CSU/DSU is used to make this conversion.

Switches A switch is another type of device used to link several LANs together. For more information refer to Chapter 12.

Summary

- LAN Interconnection Devices are the repeater, bridge, router, switch, and gateway.
- A **repeater** is used to extend the length of the network and operates at the Physical layer of an OSI model. A repeater accepts traffic from its input and repeats it at its output.
- A **bridge** is used to connect segments of same-type networks; the function of the bridge is to analyze the incoming destination frame and forward the frame to the proper segment. Bridges operate on the Data Link layer of the OSI model.
- A Learning Bridge or Transparent Bridge learns the location of each station by recording the NIC address and the port number of which frame enters the bridge.
- A Source Routing Bridge routes the frame based on information in the routing field of the frame.

- A **router** is used to route a frame from one LAN to another LAN according to its routing table. Routers operate in the Network layer of the OSI model.
- A **gateway** is used to convert one protocol to another protocol and operates in all seven layers of the OSI model.

Key Terms

Bridge

Channel Service Unit/Data Service
   Unit (CSU/DSU)

Dynamic Router

Gateway

LAN Interconnection

Learning Bridge

Repeater

Router

Source Routing Bridge

Static Router

Transparent Bridge

Review Questions

- **Multiple Choice Questions**

  1. A hub is a multiple port _____.
     - a. server
     - b. client
     - c. modem
     - d. repeater

  2. _____ operate in the Data Link layer.
     - a. Bridges
     - b. Repeaters
     - c. Switches
     - d. Gateways

  3. _____ are capable of filtering.
     - a. Bridges
     - b. Repeaters
     - c. Switches
     - d. Hubs

  4. In a _____, the frame contains the entire route to the destination.
     - a. source routing bridge
     - b. learning bridge
     - c. repeater
     - d. gateway

  5. _____ are more complex Internet working devices than bridges.
     - a. Switches
     - b. Routers
     - c. Gateways
     - d. Hubs

6. A _____ operates up to the Application layer.
   a. router   c. gateway
   b. switch   d. repeater

7. A _____ bridge learns the location of each station by recording the NIC address and the port number.
   a. source routing   c. a and b
   b. transparent   d. none of the above

8. A _____ is used to convert one protocol to another protocol.
   a. router   c. gateway
   b. switch   d. hub

● **Short Answer Questions**

1. List the LAN interconnection devices.
2. Explain the function of a repeater.
3. What is the application of a repeater?
4. The repeater operates in which layer of the OSI model?
5. Describe the function of a bridge.
6. What is the application of a bridge?
7. Which layer of the OSI model does a bridge operate in?
8. Explain the operation of a transparent bridge.
9. Explain the operation of source routing bridge.
10. List the functions of a router.
11. Explain the function of a static router.
12. When do you use a router?
13. A router works in which layer of the OSI model?
14. What is the application of a router?
15. Explain the function of dynamic routers.
16. What is the function of a gateway?
17. What is the application of a gateway?
18. A gateway operates in which layers of the OSI model?
19. What is the difference between a gateway and a router?

chapter 15

Wireless Local Area Networks (WLAN)

OBJECTIVES   After completing this chapter, you should be able to:

- Discuss the application and advantages of a Wireless LAN
- Understand Wireless LAN technology
- Describe the application of the ISM band
- Explain the operation of the Physical layer for a WLAN
- Explain the access method for WLANs

INTRODUCTION   The **Wireless Local Area Network (WLAN)** or IEEE 802.11 is a new LAN technology that enables users to access an organization's network from any location inside the organization without any physical connection to the organization's network. WLAN uses radio frequency or infrared waves as transmission signals and free air as the transmission medium. The WLAN is the next generation of campus network. Students will be able to connect their laptops to the campus network from any location inside the campus. WLAN is being used in educational institutions such as Carnegie Mellon University, making wireless access available to their students and faculty. In hospitals, WLAN facilitates doctors and nurses to access patients' files from any site in the hospital. Likewise, WLAN is used in warehouses and workshops.

15.1   Wireless LAN Topology

The simplest WLAN configuration is the **Peer-to-Peer Network**, as shown in Figure 15.1. Note in Figure 15.1 that each PC has a wireless NIC and can communicate with every other PC, as long as they are in the range of each other. Another type of configuration involves the use of a device called an **Access Point (AP)**. An access point enables the WLAN users to access the Internet and campus network.

FIGURE 15.1
Peer-to-peer
network

Antenna

Figure 15.2 shows an access point device connected to a wired LAN. The function of an AP is to receive information from clients and retransmit that information via the air and the LAN port to the hub. The AP works as a bridge between a Wireless LAN and a wired LAN.

FIGURE 15.2
WLAN with single
access point

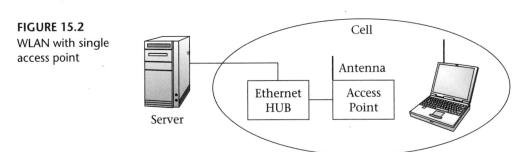

Cell

Antenna

Ethernet
HUB

Access
Point

Server

The area covered by an access point is called a **cell**. As shown in Figure 15.2, any station inside the cell can access the AP. The function of an AP is to coordinate communication between Wireless LAN clients and a wired LAN. AP provides synchronization and supports roaming. When the number of clients in a cell or the area of coverage increases, multiple access points will be added to the Wireless LAN, as shown in Figure 15.3. The access point coverage varies and depends on the manufacturer's design of the Wireless LAN product, transmitter power, the environment in which the WLAN operates (indoors or outdoors). Typically an AP can cover an area with a radius of up to 3000 feet (1000 meters) outdoors and 600 feet (200 meters) indoors.

Advantages of Wireless LAN

Following are some of the advantages of wireless LANs over wired LANs:

1. Wireless LANs can be used in places where wiring is impossible.
2. Wireless LANs can be expanded without any rewiring.
3. Wireless LANs provide the users mobility, i.e., the users can move their computers anywhere inside the organization.

FIGURE 15.3
WLAN with multiple
access points

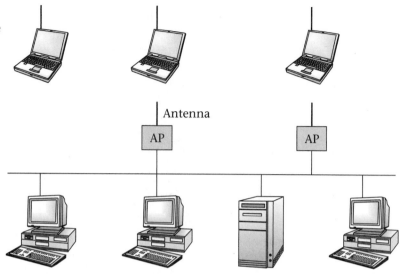

Antenna

4. Wireless LANs support roaming, users can move around with their laptops without interruption in connection.

5. Wireless LANs are cost effective as they make possible movement from one location to another without any expense for connecting wires.

15.2 Wireless LAN Technology

Two types of technology are used for transmitting information in WLANs: Infrared technology (IR), and Radio Frequency technology (RF).

Infrared Technology

IR Technology is suitable for an indoor WLAN because infrared rays cannot penetrate through walls, ceilings, or other obstacles. In **infrared** technology, the transmitter and receiver should see each other (be in the line of sight) just like a remote control for a television set. In an environment where there are obstacles, i.e., buildings, walls, and so forth, between the transmitter and a receiver, the transmitter may use diffused IR. However, most WLANs use RF technology.

Radio Frequency Technology

Movement of electrons through a conductor generates electromagnetic waves. When electromagnetic wave frequency is between 10^2 to 10^{10} Hz, it is called **Radio Frequency** (RF). This radio frequency can propagate through walls and barriers.

Characteristics of Radio Frequency

Propagation Speed: Radio waves propagate through a vacuum at the speed of light (3.0×10^8 m/s).

Frequency: Frequency is the number of cycles per second.

Wavelength: Wavelength is measured in meters and represented by the Greek character λ (lambda). It is the distance between two successive peaks of a wave or the distance traveled by one cycle of wave as shown in Figure 15.4.

FIGURE 15.4
Wavelength of sine-wave signal

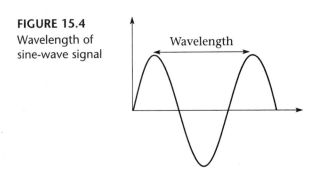

The relation between frequency and wavelength of RF signal is given by the following equation:

$$\lambda = \frac{C}{F}$$

(Equation 15.1)

where

C = Speed of Light and

F = Frequency of the Signal

Types of RF Signals   There are two types of RF signals in use for transmission of information: Narrowband signals and spread-spectrum signals.

Narrowband signal refers to the signal with a narrow spectrum as shown in Figure 15.5. In narrowband, information is transmitted at a specific frequency, such as AM and FM radio waves.

FIGURE 15.5
Narrowband signal

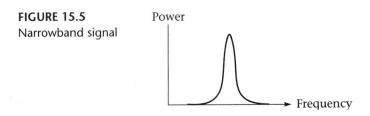

In **spread-spectrum** technology, information is transmitted over a range of frequencies as shown in Figure 15.6. Spread spectrum is one of the most popular technologies for WLAN.

FIGURE 15.6
Spread-spectrum
signal

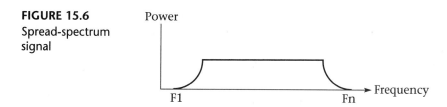

There are certain advantages to using the spread-spectrum band over the narrowband. Some of these advantages are as follows:

- In spread-spectrum technology, information is transmitted at different frequencies.
- It is hard to jam a spread-spectrum signal; the signal can not be disrupted by other signals.
- Interception of a spread-spectrum signal is more difficult than interception of a narrowband signal.
- Noise is less disruptive in spread-spectrum signal than in a narrowband signal.

ISM Band   The Federal Communication Commissions (FCC) allocates separate ranges of frequencies to radio stations , TV stations, telephone companies, and navigation and military agencies. The FCC also allocates a band of frequencies called the **Industrial, Scientific and Medical Band (ISM)** for industrial, research, and medical applications. The use of the ISM band does not require a license from the FCC (with power of transmission up to one watt). Figure 15.7 shows the ISM band.

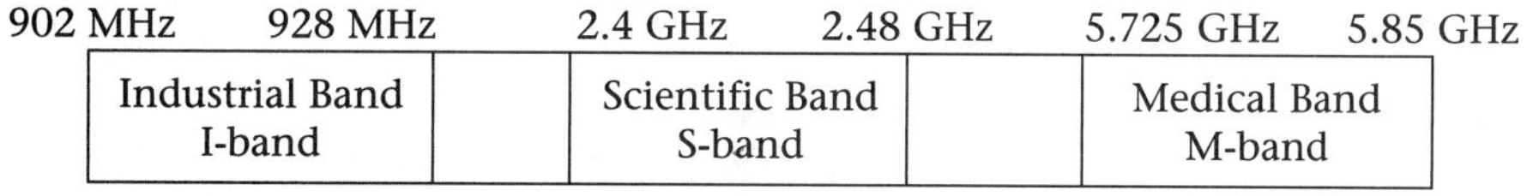

FIGURE 15.7
ISM band

902 MHz 928 MHz		2.4 GHz 2.48 GHz		5.725 GHz 5.85 GHz
Industrial Band I-band		Scientific Band S-band		Medical Band M-band

15.3   WLAN Standard (IEEE 802.11)

The IEEE 802.11 Committee has approved standards for Medium Access Control (MAC) sublayer and the Physical layer of a WLAN.

Physical Layer   In general the Physical layer of a WLAN performs following functions:

- Modulation and encoding: information is modulated then transmitted to the destination
- Supports multiple data rate

- Senses the channel to see if it is clear or not (carrier sense)
- Transmits and receives information

One of the functions of the Physical layer is to transmit information to the receiver. The IEEE 802.11 defines two types of radio frequencies (**Frequency Hopping Spread Spectrum** and **Direct Sequence Spread Spectrum**) and one infrared transmission method.

Frequency Hopping Spread Spectrum IEEE 802.11 standards recommended the scientific band (2.4 GHz to 2.483 GHz) of the ISM band for WLAN. This band is divided into 79 channels of 1 MHz each. The transmitter sends each part of its information on a different channel. Figure 15.8 shows the frequency hopping spectrum.

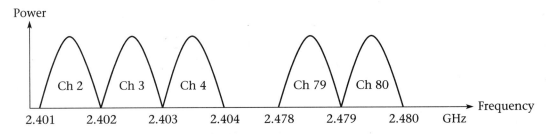

FIGURE 15.8 Frequency hopping spread spectrum

The order of channels used by the transmitter to transmit information to the receiver is predefined and the receiver knows the order of incoming channels. For example, the transmitter may use a hopping pattern of 3, 6, 5, 7, and 2 for transmitting information. The hop sequence can be selected during installation of the WLAN. The FCC requires that a transmitter spend a maximum of 400 ms in each frequency for transmission of data (this time is called Dwell time) and use 75 hop patterns (each hop is one channel). The FCC also requires the maximum power for a transmitter in the United States is one watt.

Frequency Hopping Spread Spectrum (FHSS) is more immune to the noise, because information is transmitted on different channels. In FHSS if one channel is noisy, it can retransmit information on another channel. Table 15.1 shows FHSS channels and frequencies.

Direct Sequence Spread Spectrum In **Direct Sequence Spread Spectrum (DSSS)**, before transmission each bit of information is broken into a pattern of bits called a **Chip**.

For generating chip bits, each information bit is Exclusive-ORed with **Pseudo Random Code** as shown in Figure 15.9. The output of the Exclusive-OR for each data bit are called chip bits; these chip bits are modulated and then transmitted.

TABLE 15.1 FSSH Channels and Frequencies

Channel Number	Frequency in GHz
2	2.402
3	2.403
4	2.404
—	—
—	—
80	2.48

FIGURE 15.9
Generating chips

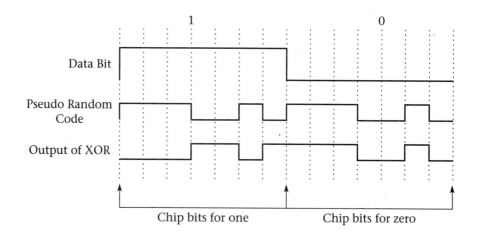

This method creates a higher modulation rate, because the transmitter transmits the chips over larger frequency spectrum. Figure 15.10 shows the transmission section of the Physical layer. The receiver uses the same pseudo random code to decode the original data. A larger chip sequence generates a larger frequency band. The IEEE 802.11 standard recommends eleven bits for each chip.

FIGURE 15.10
Transmission section
of the Physical layer

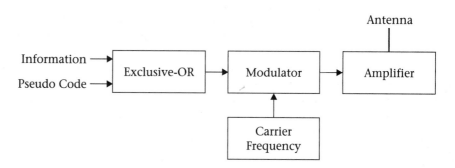

The DSSS supports two types of modulation: Differential Binary Phase Shift Keying (DBPSK), which is used for a data rate of 1 Mbps, and Differential Quadrature Phase Shift Keying (DQPSK), which is used for a data rate of 2 Mbps. The Physical layer also scrambles the information before transmitting it.

MAC Layer   The Medium Access Control (MAC) layer performs the following functions:

- Support of multiple Physical layers
- Support of access control
- Fragmentation of frame
- Frame encryption
- Roaming

IEEE 802.11 defines **Carrier Sense Multiple Access with Collision Avoidance (CSMA/CA)** as a method for a station to access the WLAN. CSMA/CA is similar to CSMA/CD.

Figure 15.11 shows two stations and one access point. Stations B and C are covered by an access point, but station B cannot cover station C. Figure 15.12 shows the CSMA/CA process:

FIGURE 15.11
CSMA/CA for a wireless LAN

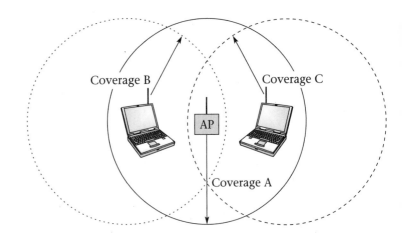

1. While the access point communicates with B, C can listen to the AP and find that the medium is not clear. Station C waits for the media to be clear. When B is transmitting to the AP, C is not in the range to detect that B is transmitting to the AP. Therefore, C will see the media as clear and start transmission. B and C will be transmitting to AP at the same time and cause a collision. Station C is a hidden station and there is no physical connection to detect this collision.

2. Station B wants to transmit to the AP. It senses the medium. If the medium is clear, it sends a short message to the AP, which is called the

FIGURE 15.12
CSMA/CA process

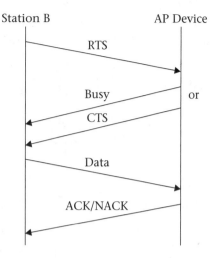

Request to Send (RTS). This message contains the destination, source address, and size of the data to be transmitted.

3. If the AP is ready to communicate with B, the AP will send a Clear to Send (CTS) frame to B, otherwise it will send a busy frame. This signal can be detected by station C and it will recognize that the medium is busy. Station B receives a CTS signal and then transmits its frame. The receiver will acknowledge each frame transmitted by B.

15.4 Characteristics of WLANs

Currently, some of the manufacturers of WLAN products offer 5.5 Mbps and 11 Mbps WLAN products by using a DSSS technique. The following section shows the characteristics of both FHSS and DSSS.

The following are characteristics of WLAN using Frequency Hopping Spread Spectrum (FHSS):

- Frequency of Operation: S-band of ISM
- Data Rate: 1 and 2 Mbps
- Modulation Technique: 2GFSK (2nd level Gaussaian Shaped FSK) and 4GFSK (Gaussian Shaped FSK using 1 million symbols per second)
- Number of Hop Channels: 79

The following are characteristics of WLAN using Direct Sequence Spread Spectrum (DSSS):

- Frequency of Operation: S-band
- Data Rate: 1 and 2 Mbps (some vendors will offer 11 Mbps products)

- Modulation technique: DBPSK and DQPSK
- Chip bits: 11 bits

Summary

- The standard for wireless LAN is IEEE 802.11.
- Components of **WLAN** are the WLAN NIC, a transmitter, an Access Point Device, and NOS.
- WLAN uses RF signals and IR signals for transmitting information.
- Infrared rays are used for indoor transmission and cannot penetrate through obstacles.
- The IEEE 802.11 defined **Frequency Hopping** and **Direct Sequencing** for Physical layers using an RF signal.
- IEEE 802.11 defined **Carrier Sense Multiple Access with Collision Avoidance (CSMA/CA)** for access methods.
- WLAN uses S-band because no license by the FCC is required.

Key Terms

Access Point (AP)

Carrier Sense Multiple Access with Collision Avoidance (CSMA/CA)

Cell

Chip

Direct Sequence Spread Spectrum

Frequence Hopping Spread Spectrum

IEEE 802.11

Industrial, Scientific and Medical Band (ISM)

Infrared (IR)

Narrowband Signal

Peer-to-Peer Network

Pseudo Random Code

Radio Frequency (RF)

Spread Spectrum

Wireless Local Area Network (WLAN)

Review Questions

- **Multiple Choice Questions**
    1. What is the transmission medium for a WLAN? _____
        - a. Air
        - b. Coaxial cable
        - c. Optical cable
        - d. UTP

2. What is the maximum transmission power for a WLAN? _____
   a. 2 watts
   c. 3 watts
   b. 1 watt
   d. 10 watts

3. What is the access method for a WLAN? _____
   a. CSMA/CD
   c. Token passing
   b. CSMA/CA
   d. Full-duplex

4. The advantage of spread-spectrum signal over a narrowband signal is _____.
   a. a spread spectrum has more power
   b. a spread spectrum signal uses a range of frequencies
   c. a narrowband signal uses a range of frequencies
   d. a spread spectrum signal using a single frequency

5. The area covered by an access point is called a _____.
   a. frame
   c. cell
   b. token
   d. chip

6. Which of following technologies are used for a WLAN? _____
   a. Infrared
   c. a and b
   b. Radio frequency
   d. none of the above

7. What is the IEEE standard for a WLAN? _____
   a. IEEE 802.10
   c. IEEE 802.12
   b. IEEE 802.11
   d. IEEE 802.13

● **Short Answer Questions**

1. What does WLAN stand for?
2. Explain narrowband signal and spread-spectrum signal.
3. What is the function of an Access Point device in a WLAN?
4. What are the advantages of spread-spectrum signals over narrowband signals?
5. What is the IEEE standard number for wireless?
6. What type of access method is used for a WLAN?
7. List the type of technologies that are used for a WLAN.
8. Explain FHSS operation.
9. What is the maximum transmitter power for a WLAN?
10. Explain access method for a WLAN.
11. What does ISM stand for?

chapter 16

Fiber Distributed Data Interface

OBJECTIVES

After completing this chapter, you should be able to:

- Describe Fiber Distributed Data Interface
- List FDDI rings and their function
- List the types of stations connected to FDDI
- Describe FDDI access method and operation
- Describe the function of a dual attachment station and a single attachment station
- Show an FDDI frame format and explain the function of each field
- Understand FDDI bit transmission

INTRODUCTION

The Fiber Distributed Data Interface (FDDI) standard was developed by the ANSI X3T9.5 Standards Committee in 1980, then submitted to the ISO. The ISO developed a new version of FDDI, which is compatible with the current ANSI standard.

16.1 FDDI Technology

Fiber Distributed Data Interface (FDDI) is a high-speed LAN using ring topology, with a data rate of 100 Mbps. FDDI uses two rings, and thus is termed a dual ring topology. In this way FDDI is similar to the IEEE 802.5 token ring. One of the most important features of FDDI is its use of optical-fiber transmission for media. The main advantage of using fiber optics over copper wiring is security, because there is no electrical signal on the media to tap.

FDDI uses two rings: one primary and one secondary ring. Traffic travels through the rings in opposite directions, as shown in Figure 16.1. The **primary ring** is used for data transmission, and the **secondary ring** is used for backup in the case of primary breakdown. FDDI allows up to 1000 stations to be connected to the ring, with a maximum ring circumference of 200 km.

FIGURE 16.1
FDDI data traffic flow

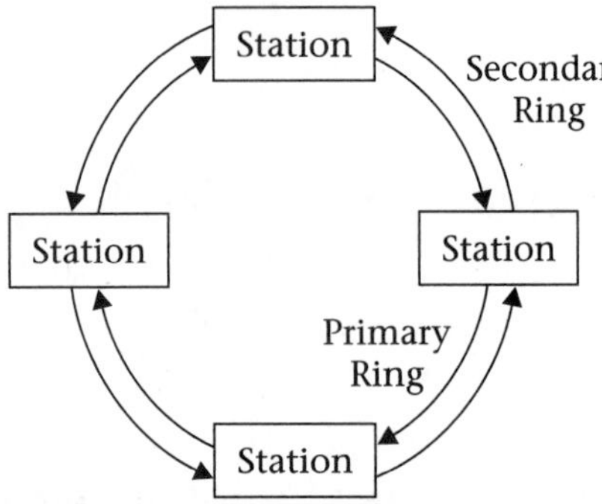

16.2 FDDI Layered Architecture

FDDI uses token passing as an access method; any station that wants to transfer information holds the token and then transmits the information. The length of time a station holds the token is called synchronous allocation time (SAT) and this time is variable for each station. The allocation of this time to each station is achieved by station management. FDDI has added a new layer to the OSI model; this new layer is called Station Management. Figure 16.2 shows FDDI layered architecture.

FIGURE 16.2
FDDI layered
architecture

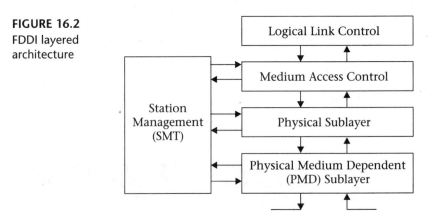

Media Access Control (MAC): Defines the medium access, frame format, addressing, token handling, and FCS calculation. FDDI also communicates with the higher layer protocols, such as TCP/IP, SNA and AppleTalk. The FDDI MAC layer accepts up to 9000 symbols from the upper layer, then adds a MAC header and passes the data to the Physical layer. The maximum frame size is 4500 bytes.

Physical Sublayer Protocol (PHY): Handles data Encoding/Decoding, framing, and clock synchronization.

Physical Medium Dependent (PMD): Defines the transmission medium, such as fiber-optic link, optical component, and connectors.

Station Management (SMT): The function of SMT is ring control, ring initialization, station insertion and station removal.

Components of FDDI   The components of FDDI are a fiber-optic cable, a **concentrator** (ring), and the stations connected to the concentrator, as shown in Figure 16.3. There are two types of stations used in FDDI.

FIGURE 16.3
Components of an
FDDI network

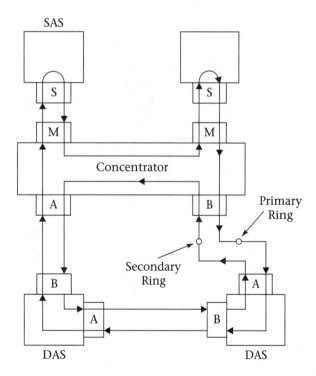

1. **Dual Attachment Station (DAS) or Class A:** DAS is attached to both rings and has two ports to connect to the ring; one connected to the primary ring and other to the secondary ring. Figure 16.4 shows the DAS ports.

2. **Single Attachment Station (SAS) or Class B:** SAS attaches to the primary ring.

FIGURE 16.4
Dual attachment
station

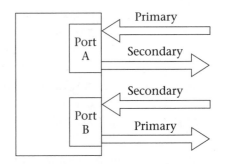

16.3   FDDI Ports

FDDI defines four types of ports in order that the Single Attachment Station (SAS) and Dual Attachment Station (DAS) can be connected to the FDDI Concentrator, thus making two rings.

The four types of ports used in FDDI topology, are shown in Figure 16.3. They are listed below.

Type A: Type A connects to the incoming primary ring and outgoing secondary ring of the FDDI dual rings. This port is used with a DAS.

Type B: Type B connects to the outgoing primary ring and incoming secondary ring of the FDDI dual rings.

Type M: Type M connects a concentrator to a single attachment station (SAS). This port is used in the concentrator.

Type S: Type S connects an SAS to a concentrator.

FDDI uses two counterclockwise rings. If one ring breaks, the outer ring will handle the traffic. If both rings break at the same point, both rings can connect together to form a single ring, as shown in Figure 16.5.

FIGURE 16.5
FDDI wrapping

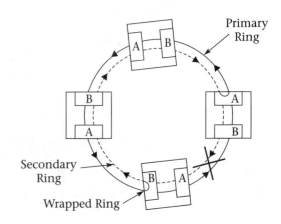

FDDI uses the principle of token ring topology. At any one time there is an active monitor in the ring to control the ring operation. FDDI uses a **4B/5B en-coding** method that converts four bits of binary into a 5-bit symbol, as shown in Table 10.1, Chapter 10.

16.4 FDDI Access Method

Any station wishing to transmit holds the token and starts a token timer, which determines how long a station can transmit. By combining a large frame size (4500 bytes) with a high transmission speed, an FDDI can transfer data very efficiently from one point to another. It is suitable for backbone networks and file servers.

16.5 FDDI Fault Tolerance

A system able to respond to unexpected hardware and software failures and continue to provide service is said to be fault tolerant.

FDDI uses two fiber-optic rings: a primary ring and a secondary ring. Traffic on these rings travels in opposite directions. The dual rings in an FDDI network provide fault tolerance for the FDDI.

If the primary ring fails, FDDI uses the secondary ring as a backup. If both rings are damaged, the dual ring automatically "wraps" (the primary ring connects to the secondary ring) and changes the ring into a single ring, as shown in Figure 16.5. Figure 16.6 shows the cable broken in two places with FDDI wrapping the ring.

FIGURE 16.6
FDDI wrapped ring

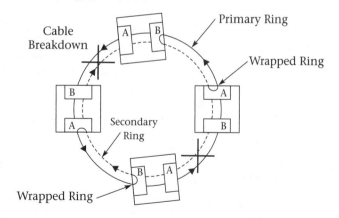

If a station on a dual ring fails or is powered down, the dual ring will wrap into a single ring. If two stations on the ring fail, the failure will cause ring segmentation. An optical relay is used to avoid segmentations of the ring by eliminating the failed station(s) from the ring. Figure 16.7 shows DAS stations connected to the rings by an optical bypass relay.

FIGURE 16.7
FDDI bit transmission

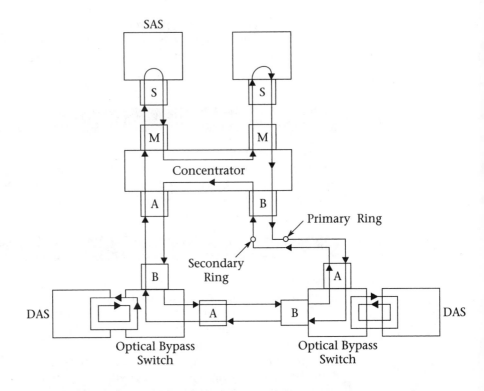

16.6 FDDI Bit Transmission

FDDI uses light pulses to transmit information from source to destination. A bit can have two values, one or zero. FDDI determines bit change by the state of the light on the receiver side. The receiver takes a sample of light every eight nanoseconds. The light is either on or off. If the light has changed since the last sample, then a one is received. If there is no change in the light, then a zero is received. Therefore, each time there is a transition of light it is translated as one (from off to on and vice versa), as depicted in Figure 16.8.

FIGURE 16.8
FDDI frame format

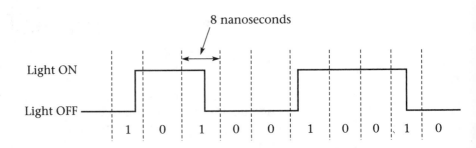

16.7 FDDI Frame and Token Formats

The FDDI frame and token formats are similar to those specified by **IEEE 802.5**. Figure 16.9 shows the FDDI frame format.

Bytes 8	1	1	2 or 6	2 or 6		4	1	1
Preamble	Start Delimiter	Frame Control	DA	SA	Data	FCS	End Delimiter	Frame Status

FIGURE 16.9 FDDI ring used for LANs backbone

The parameters for Figure 16.9 are as follows:

- Preamble (PA) consists of 16 or more IDLE symbols.
- Start Delimiter (SD) contains symbols J and K.
- Frame Control (FC) contains two symbols used to define the type of token.
- End Delimiter (ED) contains one or two control symbols (T).

16.8 FDDI Backbone

FDDI is used for LAN backbones on campuses and in corporations having several buildings in one location. Figure 16.10 shows the application of the FDDI as a backbone, using an asymmetric switch to convert from 1000 Mbps to 100 Mbps. An FDDI concentrator is used to connect two FDDI backbones together.

FDDI is one of the most expensive network backbones. With cheaper Ethernet switches and Fast Ethernet network cards, most network designers rely on a 100 Mbps Ethernet backbone rather than FDDI. Figure 16.11 shows a Fast Ethernet switch as a replacement for an FDDI backbone.

The advantages of 100 Mbps Ethernet over the FDDI are:

1. A Fast Ethernet NIC is less expensive than an FDDI NIC.
2. Fast Ethernet can use UTP as a transmission media, which is less expensive than fiber cable.

Summary

- **Fiber Distributed Data Interface (FDDI)** is a high-speed LAN with a data rate of 100 Mbps using dual ring topology and optical fiber as transmission media. FDDI is applied in network backbones.

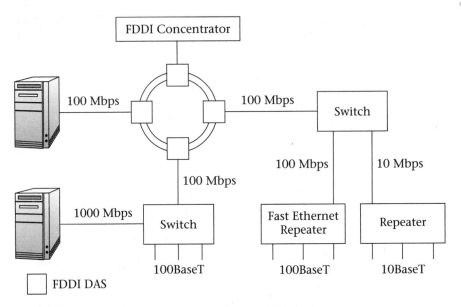

FIGURE 16.10 Fast Ethernet switch replacing FDDI ring

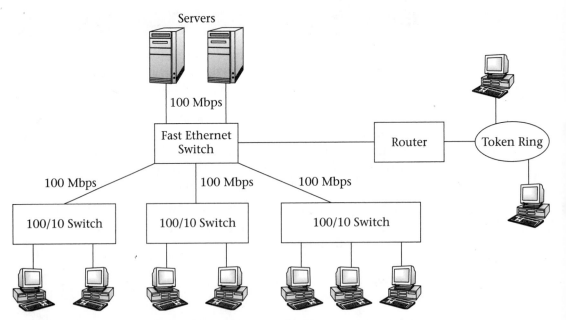

FIGURE 16.11 Fast Ethernet switch replacing FDDI backbone

- Fiber Distributed Data Interface uses two fiber-optical rings, one **primary ring** for data transmission and a **secondary ring** for back up in the case of failure of primary ring.
- FDDI uses **token passing** as its access method. The FDDI frame format is similar to the IEEE 802.5 frame format.
- A **Dual Attachment Station (DAS)** is connected to the primary and secondary rings.
- A **Single Attachment Station (SAS)** is connected only to the primary ring.
- FDDI uses **4B/5B encoding** to convert four bits to a 5-bit symbol for transmission.

Key Terms

Concentrator	Primary Ring
Dual Attachment Station (DAS)	Secondary Ring
Fiber Distributed Data Interface (FDDI)	Single Attachment Station (SAS)
4B/5B Encoding	Token Passing
IEEE 802.5	

Review Questions

- **Multiple Choice Questions**

    1. FDDI is a high-speed _____.
       a. LAN
       b. WAN
       c. a and b
       d. MAN

    2. FDDI uses _____ topology.
       a. bus
       b. star
       c. ring
       d. hybrid

    3. FDDI uses _____ rings.
       a. 2
       b. 3
       c. 4
       d. 1

    4. FDDI's transmission media is _____.
       a. UTP
       b. STP
       c. coaxial cable
       d. fiber-optic cable

5. FDDI allows up to _____ stations to be connected to the ring.
   a. 17
   b. 12
   c. 1000
   d. 500

6. The maximum frame size for an FDDI is _____.
   a. 6 bytes
   b. 64 bytes
   c. 4500 bytes
   d. 1800

7. There are _____ types of station used in FDDI.
   a. 2
   b. 3
   c. 4
   d. 5

8. DAS is connected to the _____.
   a. primary ring
   b. secondary ring
   c. a and b
   d. none of the above

9. An application of the FDDI is _____.
   a. office LAN
   b. campus backbone
   c. WAN
   d. Internet

• Short Answer Questions

1. What does FDDI stand for?
2. Why does an FDDI use two rings? What are the rings?
3. What is the data rate for FDDI?
4. Explain the FDDI access method.
5. Explain the function of concentrator.
6. Explain the function of a DAS.
7. Explain the function of a SAS.
8. What is a wrapped ring?
9. Explain the function of a bypass relay switch.
10. Show the FDDI frame format and explain the function of each field.
11. What type of encoding is used in an FDDI?
12. What are the maximum number of stations that can be connected to an FDDI network?
13. List the advantages of a Fast Ethernet switch over an FDDI.
14. What type of encoding is used in an FDDI and why?

chapter

Synchronous Optical Network (SONET)

OBJECTIVES   After completing this chapter, you should be able to:

- Describe the characteristics of a Synchronous Optical Network (SONET)
- List the components of SONET and define the function of each component
- List SONET's optical signal rates
- Show the SONET frame format and explain the function of each overhead field

INTRODUCTION   SONET is a high-speed optical carrier using fiber-optic cable for transmission media. The term SONET is used in North America and is a standard established by the American National Standards Institute (ANSI). The ITU (International Telecommunication Union) has set a standard for SONET called Synchronous Digital Hierarchy (SDH) which is also used in Europe.

SONET optical architecture is based on a four-fiber bidirectional ring to provide the highest possible level of service assurance. New application software (such as Medical Images and CAD CAM applications) require more bandwidth than other applications, and SONET provides high-speed transmission with a large bandwidth.

17.1   Characteristics of SONET

The most significant characteristics of SONET are as follows:

- SONET uses byte multiplexing at all levels.

- SONET is a high-speed transport (carrier) technology with a self-correcting path.
- SONET uses multiplexing and demultiplexing.
- SONET provides **Operation Administration and Maintenance (OAM)** functions for network managers.
- The basic electrical signal for SONET is **Synchronous Transport Signals Level One (STS-1)**.
- SONET transmits the STS-1 at the rate of 8000 frames per second.
- Slower signals can be multiplexed directly onto higher speeds.

17.2 SONET Components

Figure 17.1 shows SONET's components, which consist of an STS (Synchronous Transport Signal) multiplexer (MUX), a regenerator, an Add/Drop multiplexer, an electrical-to-optical converter, and an STS Demultiplexer. These components are described in the following list:

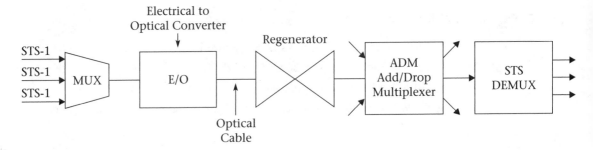

FIGURE 17.1 Shows SONET components

STS Multiplexer: The function of an STS MUX is to multiplex electrical input signals to a higher data rate and then convert the results to optical signals (as shown in Figure 17.1).

Regenerator: The regenerator performs the functions of a repeater. If the optical cable is longer than standard, the regenerator will be used to receive the optical signal and then to regenerate the optical signal.

Add/Drop Multiplexer: Add/Drop multiplexers are used for extracting or inserting lower rate signals from or into higher rate multiplexed signals without completely demultiplexing the SONET signals.

STS Demultiplexer: STS Demultiplexers convert and demultiplex optical signals to electrical signals for the users.

17.3   SONET Signal Rates

The lowest level signal in SONET is the Synchronous Transport Signal Level One (STS-1), which has a signal rate of 51.84 Mbps. The STS-1 is an electrical signal, which is converted to an optical signal called OC-1. The higher SONET data rate is represented by STS-n where n is 1, 3, 9, 12, 18, 24, 34, and 48. Table 17.1 shows SONET and SDH signal rates.

TABLE 17.1   Data Rates for OC, STS, and STM Signals

Fiber Optical (OC) Signal OC-n Level	Synchronous Transport Signal (STS) for SONET	Synchronous Transport Module (STM) for SDH	Data Rate in Mbps
OC-1	STS-1		51.84
OC-3	STS-3	STM-1	155.52
OC-9	STS-9	STM-3	446.560
OC-12	STS-12	STM-4	622.08
OC-18	STS-18	STM-6	933.12
OC-24	STS-24	STM-8	1244.16
OC-36	STS-36	STM-12	1866.23
OC-48	STS-48	STM-16	2488.32

OC = Optical Carrier Signal

STS = Synchronous Transport Signal (Electrical Signal for SONET)

STM = Synchronous Transport Module (Electrical Signal for SDH)

17.4   SONET Frame Format

A **SONET frame format** is also called a **Synchronous Payload Envelope (SPE)**. The basic transmission signal is STS-1. The STS-1 format is shown in Figure 17.2. It is made up of 9 rows and 90 columns bytes. The frame size is $90 \times 9 = 810$ bytes or $810 \times 8 = 6480$ bits. SONET transmits 8000 frames per second. The data rate for STS-1 is $6480 \times 8000 = 51.84$ Mbps.

The first 3 columns are called transport overhead, which is $3 \times 9 = 27$ bytes; 9 of these 27 bytes are used for section overhead and 18 bytes are used for line overhead. The actual data rate is 86 columns $\times$ 9 rows $\times$ 8 bits $\times$ 8000 frames/sec = 50.112 Mbps.

The STS-1 frame is transmitted by the byte from row 1, column 1 to row 9, column 90 (scanning from left to right).

FIGURE 17.2
STS-1 frame format

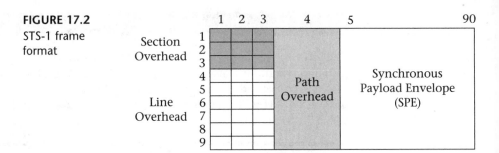

Following are descriptions of the functions of overheads in the SONET frame:

Path Overhead: Path overhead is part of SPE and contains following information: performance monitor of Synchronous Transport Signal, path trace, parity check, and path status.

Section Overhead: Section overhead contains information about frame synchronization (informing the destination of an incoming frame), carries information about Operation Administration and Maintenance (OAM), handles frame alignment, and separates data from voice.

Line Overhead: Line overhead carries the payload pointers to specify the location of SPE in the frame and provides automatic switching (for standby equipment). It separates voice channels and provides multiplexing, line maintenance, and performance monitoring.

17.5 SONET Multiplexing

Higher levels of synchronous transport signals can be generated by using byte interleave multiplexing. The STS-3 is generated by multiplexing three STS-1 signals as shown in Figure 17.3. The output of STS-3 is converted to an optical signal called OC-3. Therefore, the STS-3 frame is made of 3×90 or 270 columns and 9 rows with the total of 2430 bytes. The STS-3 is transmitted at 8000 frames per second, therefore, the data rate of STS-3 is:

$$2430 \text{ bytes} \times 8 \text{ bits} \times 8000 \text{ frames/sec} = 155.52 \text{ Mbps}$$

FIGURE 17.3
Multiplexing three STS-1s to generate STS-3

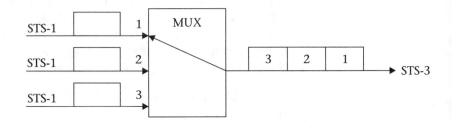

Figure 17.4 shows the STS-3 frame format. The transport overhead is made up of 9 columns and 9 rows, the STS-3 path overhead is one column, and the SONET payload envelope is 260 × 9 bytes. The STS-9 is generated by multiplexing three STS-3s, as shown in Figure 17.5.

FIGURE 17.4
Frame format of STS-3

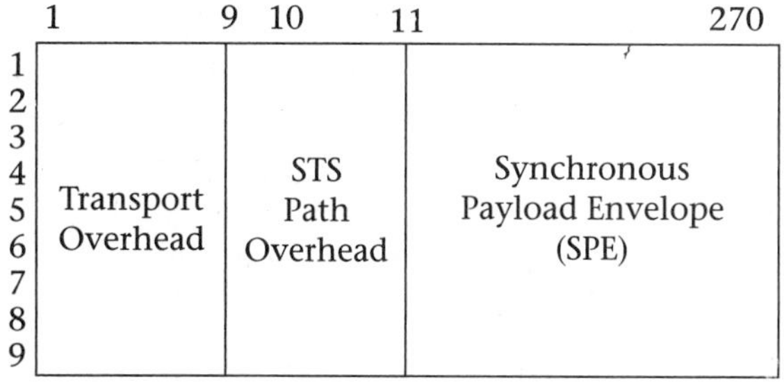

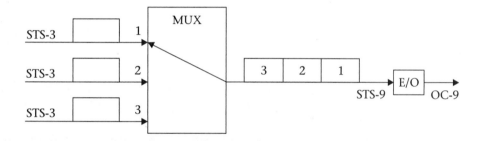

FIGURE 17.5
Generating STS-9

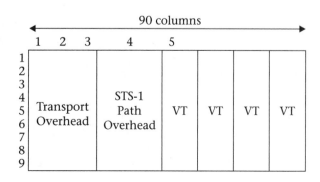

17.6 Virtual Tributaries

The basic frame for the SONET is STS-1 with a data rate of 51.84 Mbps. The payload of STS-1 is made up of 86 columns and 9 rows. In order for the SONET to carry lower data rate frames such as DS-1 and DS-2, the lower data rate frame is mapped to the STS-1 payload and is called a **Virtual Tributary (VT)**. Figure 17.6 shows the VTs in the STS-1 payload.

FIGURE 17.6
Virtual tributaries

There are four types of VTs that map into the STS-1 payload.

- VT1.5 is a frame of 27 bytes that is made up of 3 columns and 9 rows, as shown in Figure 17.7. The data rate of VT1.5 is calculated as follows:

Data Rate VT1.5 = 27 bytes × 8 bits × 8000 frames/sec = 1.728 Mbps

The VT1.5 is used for transmission of DS-1 with a data rate of 1.54 Mbps. The STS-1 payload can transmit 28 VT1.5s.

FIGURE 17.7
VT1.5 mapped into STS-1 frame

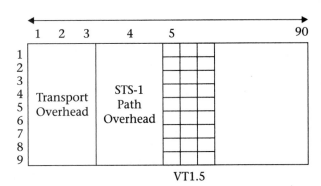

VT1.5

- VT2 is a frame of 36 bytes (made up of 4 columns and 9 rows) that is used for transmission of an European E-1 line (an E-line can carry 30 voice channels) with a data rate of 2.3048 Mbps.
- VT3 is a 54-byte frame (made up of 6 columns and 9 rows) that is used for transmission of a DS-1C frame with a data rate of 3.152 Mbps.
- VT6 is a 108-byte frame (made up of 12 columns and 9 rows) that is used for transmission of a DS-2 frame with a data rate of 6.312 Mbps.

Summary

- The Synchronous Optical Network (SONET) is a high-speed data carrier. The term SONET is used in North America, and Synchronous Digital Hierarchy (SDH) is used in Europe.
- SONET uses fiber-optical cable for transmission media, since optical transmission is immune to the interface and can transmit data over a long distance.
- SONET converts a **Synchronous Transport Signal (STS-1)** to an optical signal. The optical signal is called an **OC-1 (Optical Carrier)** and is transmitted at the rate of 8000 frames per second.
- SONET components are an STS Multiplexer, a regenerator, an Add/Drop multiplexer (ADM), and an STS demultiplexer.
- The basic transmission signal for SONET is STS-1. The frame format of SONET is made up of 9 rows and 90 columns of bytes. The frame size is 810 bytes and

this frame is transmitted at a rate of 8000 frames per second, which gives a data rate of 51.84 Mbps for STS-1.

- Three STS-1 signals are multiplexed and converted to optical to generate an OC-3 with a data rate of 155.52 Mbps.

Key Terms

Add/Drop Multiplexer	STS Demultiplexer
Line Overhead	STS Multiplexer
Operation Administration and Maintenance (QAN)	Synchronous Optical Network (SONET)
Optical Carrier Signal (OC)	Synchronous Payload Envelope (SPE)
Path Overhead	Synchronous Transport Signal (STS)
Regenerator	Synchronous Transport Signal Level One (STS-1)
Section Overhead	Virtual Tributary (VT)
SONET Frame Format	

Review Questions

- **Multiple Choice Questions**

    1. _____ sets the standard for SONET.
       a. IEEE
       b. ANSI
       c. ITU
       d. ISO

    2. SONET uses byte multiplexing in _____ levels.
       a. upper
       b. mid
       c. all
       d. none of the above

    3. The basic electrical signal for SONET is _____.
       a. STS-1
       b. STS-2
       c. STS-3
       d. STS-n

    4. SONET transmits the STS-1 at a rate of _____ frames/second.
       a. 6000
       b. 7000
       c. 8000
       d. 1000

    5. The _____ performs the function of a repeater.
       a. regenerator
       b. STS multiplexer
       c. STS demultiplexer
       d. SONET

6. STS-1 has data rate of _____ Mbps.
   a. 810
   b. 8000
   c. 51.84
   d. 1.54

7. SONET is a/an _____.
   a. LAN
   b. WAN
   c. Optical Carrier
   d. Internet

8. STS-1 frame is made up of _____.
   a. 9 columns and 90 rows
   b. 9 rows and 90 columns
   c. 10 rows and 100 columns
   d. none of the above

9. The optical signal for STS-1 is _____.
   a. OC-3
   b. OC-1
   c. OC-2
   d. OC-n

10. STS-3 is generated by multiplexing _____.
    a. three STS-1s
    b. six STS-1s
    c. five STS-1s
    d. three STS-3s

11. STS-9 is generated by multiplexing _____.
    a. six STS-1s
    b. three STS-3s
    c. three STS-1s
    d. two STS-3s

12. An STS-3 frame format is made up of _____.
    a. 270 columns and 9 rows
    b. 9 rows and 270 columns
    c. 10 rows and 300 columns
    d. none of the above

13. SONET uses Virtual Tributary to transmit _____.
    a. a frame with a data rate higher than STS-1
    b. a frame with a data rate less than STS-1
    c. a frame with an STS-1 data rate
    d. none of the above

• Short Answer Questions

1. What does SONET stands for?
2. What does SDH stands for?
3. What is an application of SONET?
4. What is the basic electrical signal for SONET?
5. What are transmission media for SONET?

6. List some of the advantages of SONET.

7. List the SONET components.

8. What does STS-1 stand for?

9. What is OC-1?

10. What is the data rate for STS-1?

11. How many bytes is STS-1?

12. How many STS-1s must be multiplexed to generate an STS-3?

13. SONET transmits how many frames per second?

14. Show the SONET frame format.

15. Calculate the the VT2 data rate.

16. Calculate the VT3 and VT6 data rates.

17. What is the function of the payload pointer?

18. Explain the function of Add/Drop multiplexing.

19. What is STS-n?

20. Why is STS-1 data rate 51.84 Mbps?

21. What is the function of each of the following fields in SONET frame format?

    a. Payload Pointer   c. Line Overhead

    b. Path Overhead   d. Section Overhead

chapter 18

Frame Relay

OBJECTIVES

After completing this chapter, you should be able to:

- Discuss the application of a frame relay network
- List the components of a frame relay network
- Describe the function of a frame relay switch
- Describe the function of a frame assembler/disassembler
- Show the frame relay format and explain the function of each field

INTRODUCTION

Frame relay was originally designed for use over ISDNs. The initial proposal was submitted to ITU-T for standardization in 1984. The ITU-T ratified this proposal and this standard specification adds relay and routing functions to the Data Link layer of the OSI model.

Large organizations and corporations have multiple sites in different locations and they need to connect LANs together. One solution is to lease data communication lines and connect their LANs together. A corporation with 100 offices in different locations in a country must lease 100 lines to connect their LANs, and this method is not cost effective. The **frame relay** is a network that offers frame relay services to the corporations, thus enabling their LANs to communicate with each other regardless of varying protocols. The public carriers offer frame relay networking which is less expensive than leased lines.

Frame relay is a packet-switching protocol service offered by telephone corporations to replace the X.25 protocol. It is a Wide Area Network (WAN). Figure 18.1 shows the architecture of a frame relay consisting of a **Frame Relay Network**, a **Frame Relay Assembler/Disassembler (FRAD)** and user LANs. The function of the FRAD (also called a router) is to convert the frame format from a site network to the frame format of the frame relay network and vice versa.

FIGURE 18.1
Frame relay
architecture

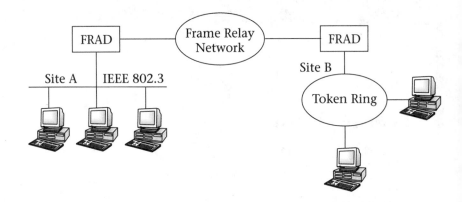

Figure 18.2 shows a reference model for a frame relay. The frame relay operates in the Data Link layer and Physical layer of the OSI model.

FIGURE 18.2
Frame relay network
reference model

Data Link Layer
Physical Layer

18.1 Frame Relay Network

Figure 18.3 shows a frame relay network, which consists of frame relay switches. The switches are connected to each other by T-1 or T-3 links, or by optical cables. These switches are located at the central office of each public carriers and the frame relay networks controlled by telephone corporations.

FIGURE 18.3
Frame relay network

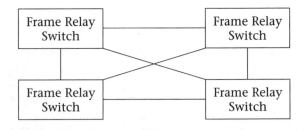

18.2 Components of Frame Relay

Frame relay is a service provided by public carriers to organizations in order that the organization can connect their networks which are located in various locations, together. For a network to be connected to a frame relay network, the following components are required:

- Frame relay access equipment: (1) Customer premises equipment such as a LAN, and (2) Access equipment, which is a Frame Relay Assembler/ Disassembler (FRAD) such as a router
- Frame relay switches
- Frame relay service: Public carriers provide frame relay services.

Function of Frame Relay Switch

The **frame relay switch** is located in the central office of a telephone company and performs the following functions:

- The switch checks for errors in a packet received from the user or previous switch. If the packet contains an error then the switch discards the packet, otherwise it transfers the packet to the next switch or destination.
- Frame relay switches use statistical packet multiplexing for multiplexing and routing.
- The end user handles data integrity such as sequencing, error detection, and retransmission.
- When a packet is transmitted between switches, there is no acknowledgment between switches.

18.3 Frame Relay Frame Format

Figure 18.4 shows the frame format of a frame relay that consists of a starting flag (1 byte), an address (2 to 4 bytes), an information field (up to 4096 bytes), a frame check sequence (FCS) generated by a 16-bit CRC, and an ending flag (1 byte).

The following are descriptions of the function of each field in the Frame Relay frame format:

Starting Flag (SF) and **Ending Flag (EF):** Set to 0111110.

Data Link Connection Identifier (DLCI): The DLCI is 10 bits. This field determines the PVC (Permanent Virtual Circuit) value, which is used by the frame relay switch to find a path for the frame to reach its destination.

Command/Response (C/R): This field determines whether the frame is a command frame or whether it is a response to a command.

Extended Address (EA): This bit identifies whether there is an extension address in the frame format.

EA = 0 means another address byte follows the current address byte.

EA = 1 means this is the last byte of the address field.

D/C: This bit determines whether the low order bits of the DLCI are control bits or DLCI bits.

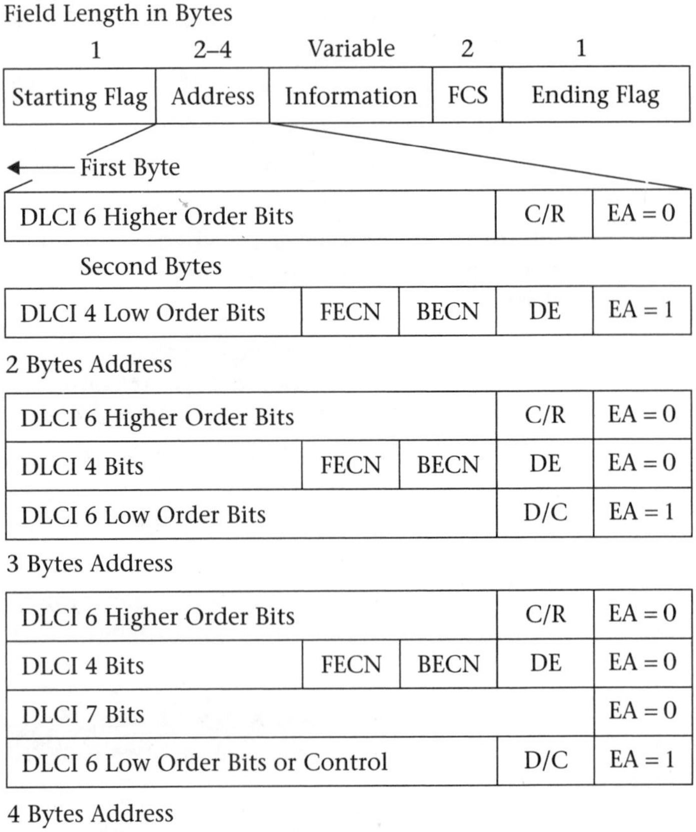

FIGURE 18.4
Frame format of
frame relay

Field Length in Bytes

1	2–4	Variable	2	1
Starting Flag	Address	Information	FCS	Ending Flag

First Byte

DLCI 6 Higher Order Bits		C/R	EA = 0

Second Bytes

DLCI 4 Low Order Bits	FECN	BECN	DE	EA = 1

2 Bytes Address

DLCI 6 Higher Order Bits			C/R	EA = 0
DLCI 4 Bits	FECN	BECN	DE	EA = 0
DLCI 6 Low Order Bits			D/C	EA = 1

3 Bytes Address

DLCI 6 Higher Order Bits			C/R	EA = 0
DLCI 4 Bits	FECN	BECN	DE	EA = 0
DLCI 7 Bits				EA = 0
DLCI 6 Low Order Bits or Control			D/C	EA = 1

4 Bytes Address

Congestion Control: The following fields are used for congestion control.

- **Forward Explicit Congestion Notification (FECN):** This bit is used to notify the end user of congestion. It can be set by switches to show that there is congestion in the direction the frame is traveling.
- **Backward Explicit Congestion Notification (BENC):** This bit can be set by switches to show that there is congestion in the opposite direction from the one the frame is traveling.

Discard eligibility (DE): Since the congestion switch might discard some frames, this bit is set to one, which instructs the switch not to discard the frame.

18.4 Frame Relay Operation

Figure 18.5 shows a frame relay network. Network A is connected to switch A through FRAD and Network B is connected through a router (Frame Relay Access

FIGURE 18.5
Frame relay network
with three switches

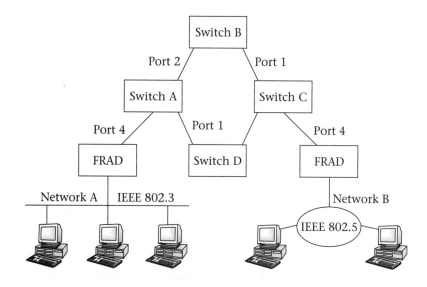

Device) to switch C. Each connection to the switch has a unique DLCI assigned by the service provider to each PVC. The path defined between two sites is called a **virtual circuit**. Frame relay supports **permanent virtual circuit (PVC)** and **switched virtual circuit (SVC)**. Frame relay supports multiple PVCs simultaneously, therefore frame relay can connect multiple sites with a single connection.

For Network A to communicate with B, a permanent virtual connection is set up by the frame relay network administrator. Each switch has a table, which indicates the incoming frame DLCI, and an output port as shown in Table 18.1.

TABLE 18.1 Routing Table for Switches A, B, and C in Figure 18.5

Switch A		Switch B		Switch C	
DLCI	Port #	DLCI	Port #	DLCI	Port #
45	2	45	1	45	4

The following example tracks how a packet is transmitted from Network A to Network B using Figure 18.5 and routing Table 18.1. Network A transmits a packet to Network B. The following steps take place for the packet to successfully reach Network B:

1. Network A transmits the packet to switch A.
2. Switch A checks for errors in the packet; if there is an error in the packet, the switch discards the packet. If the packet is correct, switch A uses its routing table (Table 18.1) to find out which port (based on DLCI 45) the packet has to exit.

3. The packet exits from port #2.

4. The packet reaches switch B.

5. Switch B checks for errors and transmits the packet, based on DLCI 45, to port #1.

6. The packet reaches switch C and switch C checks for errors. Then, using its routing table based on DLCI 45, switch C transmits the packet to port #4.

7. The packet reaches FRAD. FRAD changes the frame format of the packet from frame relay format to token ring format and transmits the frame to the token ring network.

Summary

- The application of a **frame relay** is to connect the networks of a corporation having several offices at different locations.
- Frame Relay operates at the Data Link layer of the OSI model.
- A frame relay network consists of frame relay switches, which are connected together by a T-1 or T-3 link.
- A frame relay is a service offered by a telephone company.
- The components of a frame relay are a customer LAN, a frame relay assemler/disassembler (FRAD), and a frame relay switch.
- The function of a **FRAD** is to convert a customer's LAN frame format to the frame format of a frame relay and vice versa.
- A **frame relay switch** is located in the central office of a telephone company. The function of a switch is to accept a frame from FRAD and transmit based on Data Link Connection Identifier (DLCI), to the next switch.

Key Terms

Frame Relay	Frame Relay Switch
Frame Relay Assembler/Disassembler (FRAD)	Permanent Virtual Circuit (PVC)
	Switched Virtual Circuit (SVC)
Frame Relay Network	Virtual Circuit

Review Questions

- ### Multiple Choice Questions

  1. Frame relay was originally designed for use over a/an _____ network.

     a. Internet

     b. SONET

     c. ISDN

     d. WAN

2. Frame relay network works in the _____ layer of OSI model.
   a. Data Link
   b. Transport
   c. Session
   d. Physical

3. The frame relay switches are located _____.
   a. in a user's home
   b. at the central office of a telephone company
   c. at a government office
   d. at a corporation

4. EA = _____ means this is the last byte of the address field.
   a. 0
   b. 1
   c. 2
   d. 3

5. Frame relay is used as a _____.
   a. LAN
   b. WAN
   c. MAN
   d. Internet

6. Frame relay service is provided by _____.
   a. Internet service providers
   b. telephone corporations
   c. the U.S. government
   d. none of the above

7. A LAN can be connected to a frame relay network by a _____.
   a. repeater
   b. switch
   c. FRAD
   d. gateway

8. What type of connection is used by frame relay? _____
   a. Permanent virtual circuit
   b. Virtual circuit
   c. Circuit switching
   d. Message switching

9. Frame relay is used to connect _____.
   a. the LANs that are located in different cities
   b. the LANs that are located on a campus
   c. the LANs that are located in a city
   d. the LANs that are located in one room

● Short Answer Questions

1. What is the application of frame relay?
2. Explain the function of the FRAD.

3. Explain the function of the frame relay switch.

4. What are the components of frame relay networks?

5. Explain the function of the following fields in the frame relay frame format.
   a. DLCI
   b. EA
   c. C/R
   d. DE
   e. FECN and BECN

chapter 19

Internet Architecture and TCP/IP

OBJECTIVES   After completing this chapter, you should be able to:

- Discuss the history of the Internet
- List the applications of the Internet and explain the function of each application protocol
- Explain the functions of the Internet Architecture Board (IAB)
- List Transmission Control Protocols and Internet Protocols (TCP/IP) and describe the service of each protocol
- Distinguish between IP address classes and understand how IP addresses are assigned to a network of an organization
- Describe the Domain Name Server (DNS)
- Show TCP/IP reference model
- Show User Datagram Protocol (UDP) packet format and define the function of each field
- List Applications Protocol for Transmission Control Protocol (TCP)

- Describe the function of TCP, show the TCP packet format, and describe the function of each field
- Explain the function of Internet Protocol (IP) and identify IP packet format
- Describe the application of Point-to-Point Protocol (PPP) and its packet format
- Explain TCP connection and disconnection
- Show the IPv6 format and explain the function of each field
- Describe the advantages of IPv6
- Describe Internet II

INTRODUCTION   The term Internet, short for Internetwork, describes a collection of networks that use the TCP/IP protocol to communicate among nodes. These networks are connected together through routers and gateways. Figure 19.1 shows an organization whose networks are connected together by an internal gateway that is, in turn, connected to the external gateway of the Internet. Today the term **gateway** defines a process that connects two different networks that have different protocols.

FIGURE 19.1
Connection of a
network to the
Internet

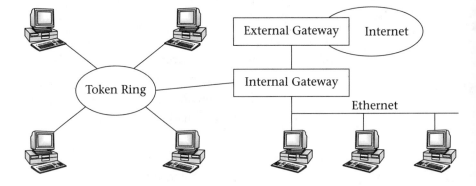

In 1968 the United States Department of Defense (DOD) created the Defense Advanced Research Project Agency (DARPA) for research on packet-switching networks. In 1969, DARPA created the **Advanced Research Project Agency (ARPA)**. In the same year ARPA selected Bolt, Beranek and Newman (BBN), a research firm in Cambridge, Massachusetts, to build an experimental network (ARPANET) to provide a test bed for emerging network technology. **ARPANET** originally connected four nodes, the University of California at Los Angeles (UCLA), the University of California at Berkeley (UCB), Stanford Research Institute (SRI), and the University of Utah to share information and resources across long distances. ARPANET experienced rapid growth with the addition of universities. At that time, the protocol used in ARPANET was called the network control protocol (NCP). NCP did not scale well to the growing ARPANET, and in 1974 TCP/IP was introduced.

In 1980 the TCP/IP protocol became the only protocol that was in use on ARPANET. At the same time most universities were using the UNIX operating system, which was created by Bell Labs in 1969. The University of California at Berkeley integrated the TCP/IP protocol into Version 4.1 of its software distribution, later known as the Berkeley Software Distribution (Berkeley UNIX or BSD UNIX). The DOD separated the military network (MILNET) from the nonmilitary network (ARPANET).

In 1985 the National Science Foundation (NSF) connected the six supercomputer centers together and named this network **NSFNET**. The NSFNET was then connected to ARPANET. Naturally, NSF encouraged universities to connect to NSFNET. Due to the growth of NSFNET, in 1987 NSF accepted a joint proposal from IBM, MCI Corporation and MERT Corporation to expand the NSF backbone. Figure 19.2 shows the NSF backbone in 1993. By 1995 numerous companies were running commercial networks.

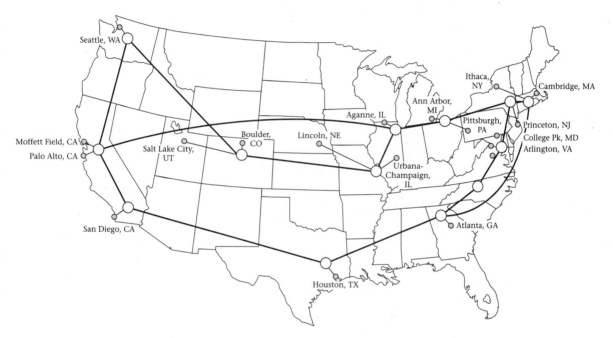

FIGURE 19.2 NSFNET backbone

The current Internet backbone is a connection of several backbones that belong to Internet network service providers such as MCI, AT&T, IBM, Sprint, and GTE. These backbones are connected through gateways. Figure 19.3 shows the GTE Internet backbone.

Internet Address Assignment Any organization wishing to connect its network to the Internet must contact the Internet Network Information Center (InterNIC) to obtain an Internet address (IP address). The following are Internet Network Information addresses:

www.rs.internic.net

E-mail: Hostmaster@internic.net

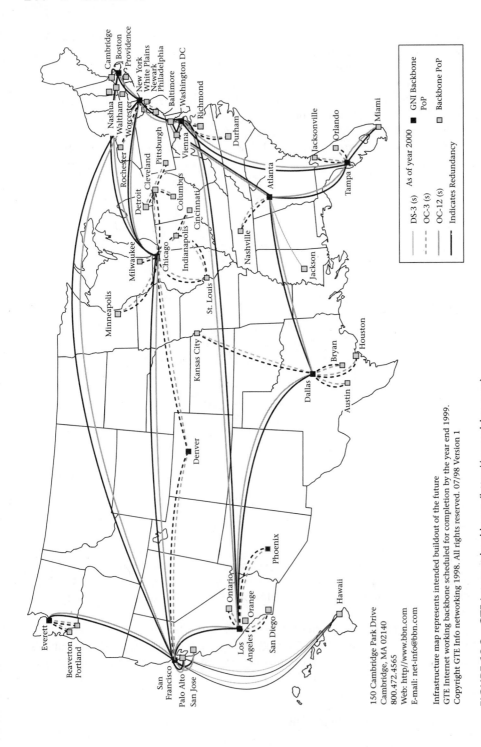

150 Cambridge Park Drive
Cambridge, MA 02140
800.472.4565
Web: http//www.bbn.com
E-mail: net-info@bbn.com

Infrastructure map represents intended buildout of the future
GTE Internet working backbone scheduled for completion by the year end 1999.
Copyright GTE Info networking 1998. All rights reserved. 07/98 Version 1

FIGURE 19.3 GTE Internet backbone (http://www.bbn.com)

Mailing address: Network Information Center
333 International
Menlo Park, CA 94025

Any organization that obtains a network IP address will submit its server's name to InterNIC. InterNIC will ensure that no two servers have the same name.

For example: **elahi@scsu.ctstateu.edu** is the author's Internet address. Reading from right to left the above domain name

edu: represents that Elahi is at some U.S. educational site

ctstateu: represents Connecticut State University

scsu: represents the machine that has information about Elahi's IP address.

19.1 The Internet Architecture Board (IAB)

The IAB comprises thirteen members; six of them are selected by the Internet Engineering Task Force (IETF). The functions of the IAB are:

- Determination of the future of Internet addressing
- Architecture of the Internet
- Direction of IETF: management of a top level Domain Name System

The following is a list of the subcommittees of the IAB and their functions:

A. **IESG:** The Internet Engineering Steering Group (IESG) works on Internet standards and also oversees the work of all the other groups.

B. **IETF:** The Internet Engineering Task Force (IETF) is an open international committee of network designers, vendors and researchers. The IETF is divided into subgroups, which are organized based on their area of expertise, such as routing or transport.

C. **IRTF:** The function of the Internet Research Task Force (IRTF) is to promote long- and short-term research related to Internet protocols such as TCP/IP, Internet Architecture, and IPv6.

D. **IANA:** The Internet Assigned Numbers Authority (IANA) works under the Internet Network Information Center (INIC). InterNIC consists of Network Solutions Inc. and AT&T Corp. The function the INIC is registration and education services. InterNIC manages registration of the second level domain names under the following top level domains: **gov, com, net, mil**, and **org**.

19.2 TCP/IP Reference Model

Transmission Control Protocol and Internet Protocol (TCP/IP) essentially consists of four levels: the Application Level, Transport Level, Internet Level, and Network Level as shown in Figure 19.4. Table 19.1 shows TCP/IP protocols and their functions.

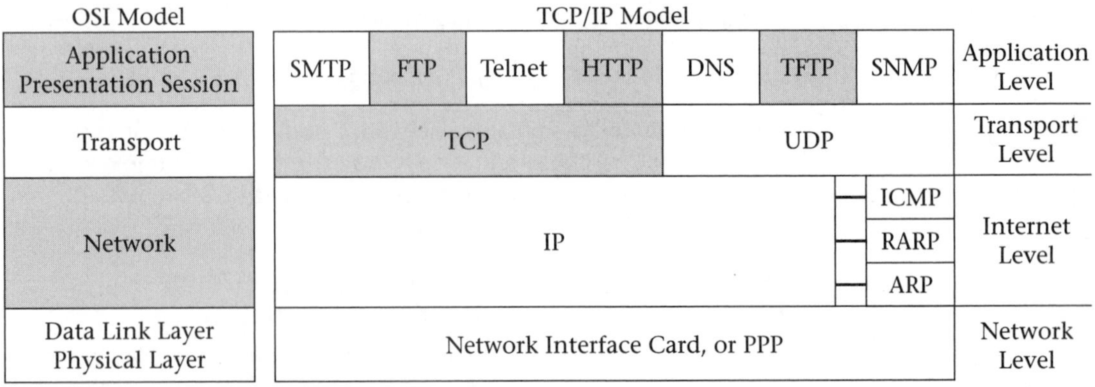

OSI Model	TCP/IP Model							
Application Presentation Session	SMTP	FTP	Telnet	HTTP	DNS	TFTP	SNMP	Application Level
Transport	TCP				UDP			Transport Level
Network	IP						ICMP RARP ARP	Internet Level
Data Link Layer Physical Layer	Network Interface Card, or PPP							Network Level

FIGURE 19.4 TCP/IP reference model

TABLE 19.1 TCP/IP Protocols and Their Functions

Protocol	Service
Internet Protocol IP	Provides packed delivery between networks
Internet Control Message Protocol ICMP	Controls transmission errors and controls messages between hosts and gateways
Address Resolution Protocol ARP	Requests physical address from source
Reverse Address Resolution Protocol RARP	Response to the ARP
User Datagram Protocol UDP	Provides unreliable service between hosts (transfer of data without acknowledgment)
Transmission Control Protocol TCP	Provides reliable service between hosts
Simple Network Management Protocol SNMP	Used for diagnostics purposes between hosts

19.3 TCP/IP Application Level

The Application level enables the user to access the Internet. Following are some of the Internet applications:

- Simple Mail Transfer Protocol (SMTP): E-mail
- Telnet
- File Transfer Protocol (FTP)
- Hypertext Transfer Protocol (HTTP)
- Simple Network Management Protocol (SNMP)
- Domain Name System (DNS)

Simple Mail Transfer Protocol

SMPT is used for **E-mail** (electronic mail); it is used for transferring messages between two hosts. To send a mail message, the sender types in the address of the recipient and a message. The electronic mail application accepts the message/mail (if the address is right) and deposits it in the storage area/mailbox of the recipient. The recipient then retrieves the message from his/her mailbox.

What Is an E-mail Address?   An E-mail address (or just "e-mail") is made up of a **Username @ Mail Server Address**. For example: In **Rizzo@scsu.ctstateu.edu**, "Rizzo" is the Username and "scsu.ctstateu.edu" is the domain name of the mail server. The letters "scsu" stands for Southern Connecticut State University, "ctstateu" stands for Connecticut State University, and "edu" stands for education.

Some e-mail addresses are little more complicated, e.g., look at **Uppal@scus1. scsu.ctstateu.edu** where "Uppal" is the Username, "scus1" is the name of a workstation that is a part of "scsu," which is located at "ctstateu," and "edu" tells you that it's an education center.

Telnet or Remote Login

Telnet is one of the most important Internet applications. It enables one computer to establish a connection to another computer. Users can login to a local computer and then remote login across the network to any other host. The computer establishing the connection is referred to as the local computer; the computer accepting the connection is referred to as the remote or host computer. The remote computer could be a hardwired terminal or a computer in another country. Once connected, the commands typed in by the user are executed on the remote computer. What the user sees on his/her monitor is what is taking place on the remote computer.

Remote login was developed for Berkeley UNIX to work with UNIX systems only, but it has been ported to other operating systems also. Telnet uses the client/server model. That is, a local computer uses a Telnet client program to establish the connection. The remote or host computer runs the Telnet server version to accept the connection and sends responses to requests.

File Transfer Protocol

FTP is an Internet standard for file transfer. It allows Internet users to transfer files from remote computers without having to log into them. FTP establishes a connection to a specified remote computer using an FTP remote-host address. Once connected, the remote host will ask the user for identification and a password. Upon compliance, the user can download or upload files.

Some sites make files available to the public. To access these files, users can enter *anonymous* or *guest* for identification and use his/her Internet address as a password. This application is called **anonymous FTP**.

Hypertext Transfer Protocol

HTTP is an advanced file retrieval program that can access distributed and linked documents on the web. Messages in HTTP are divided into request and response categories and work on the client/server principle. The request command is sent from the client to the server. The response command is sent from the server to client.

HTTP is a stateless protocol that treats each transaction independently. A connection is established between a client and a server for each transaction and is terminated as soon as the transaction is complete.

Simple Network Management Protocol

SNMP is used by the network administrator to detect problems in the network, such as router and gateway. SNMP provide information for monitoring and controlling the network. SNMP is divided to the two parts: SNMP management system and SNMP agent. The management system issues a command to the SNMP agent, and the SNMP agent responds to the command. An SNMP management system can manage network devices remotely.

Domain Name System

A **domain name** is a unique name used to identify and locate computers connected to the Internet, e.g., "scsu.ctstateu.edu." DNS is a collection of databases containing information about domain names and corresponding IP addresses. Top level domain names can be categorized as shown in Figure 19.5. A few are also listed below.

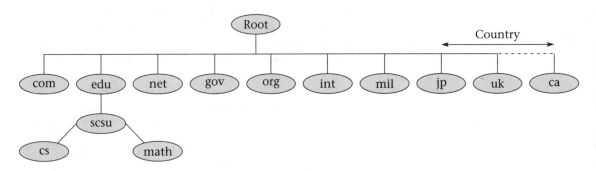

FIGURE 19.5 DNS top level domain names

EDU	Educational sites	**MIL**	Military sites
GOV	Government sites	**ORG**	Organization (non-profit) sites
COM	Commercial sites	**INT**	International organization sites

Country codes are used to identify international sites. A two letter abbreviation can be used for a particular country such as "uk" for United Kingdom, or "fr" for France.

19.4 Transport Level Protocols: UDP and TCP

The Transport layer of the TCP/IP protocol consists of **UDP (User Datagram Protocol)** and **TCP (Transmission Control Protocol). The UDP protocol** performs an unreliable connection service for receiving and transmitting data. TCP performs reliable delivery of data. TCP adds a sequence number to each packet. When a packet reaches its destination, the destination acknowledges the sequence number of the next packet it expects to receive.

User Datagram Protocol   User Datagram Protocol (UDP) applications are Trivial File Transfer Protocol (TFTP) and Remote Call Procedure (RCP). UDP accepts information from the Application layer and adds a source port, destination port, UDP length, and UDP checksum. The resulting packet is called UDP datagram packet. The total header of a UDP datagram is eight bytes. UDP passes the UDP packet to the IP protocol. The IP protocol adds its header to the packet and passes the packet to the Logical Link Control (LLC). The LLC generates a 802.2 frame (LLC frame) and passes the LLC frame to the Medium Access Control (MAC) layer, which adds its own header and transfers the frame to the Physical layer for transmission (as shown in Figure 19.6).

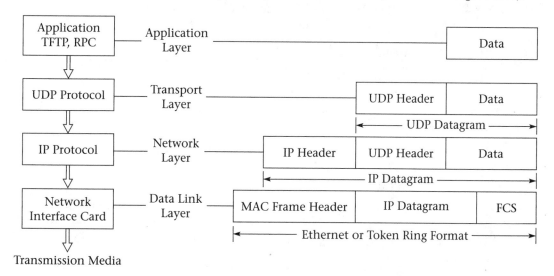

FIGURE 19.6   UDP operation

The UDP allows applications to exchange individual packets over a network as datagrams. A UDP packet sends information to the IP protocol for delivery. There is no guaranteed reliability. Figure 19.7 shows the UDP packet format.

0	31
Source Port 16 bits Defines application TFTP is port 69	**Destination Port** 16 bits Specifies destination port on server
UDP Length 16 bits Defines number of bytes in UDP header and data	**Checksum** 16 bits Checksum used for error detection of UDP header and data
Data	

FIGURE 19.7   UDP packet format

Transmission Control Protocol   Most applications prefer to use reliable delivery of information. TCP offers reliable delivery of information through the Internet. TCP gives the user a method to transmit data in a reliable fashion. In TCP, before data is transmitted to the destination, a connection (not a physical connection) must be established before the information is transmitted. TCP assigns a sequence number to each packet. The receiving end checks the sequence number of all packets to ensure that they are received. When the receiving end gets a packet, it responds to the destination by acknowledging the next sequence number. If the sending node does not receive an acknowledgment within a given period of time, it retransmits the previous packet.

Figure 19.8 shows application data passing through TCP. TCP adds a 20-byte header and passes it to IP. IP adds its header and passes it to a Network Interface Card (NIC). The NIC adds a MAC header to the information and transmits the packet. Figure 19.9 on the next page shows the TCP packet format.

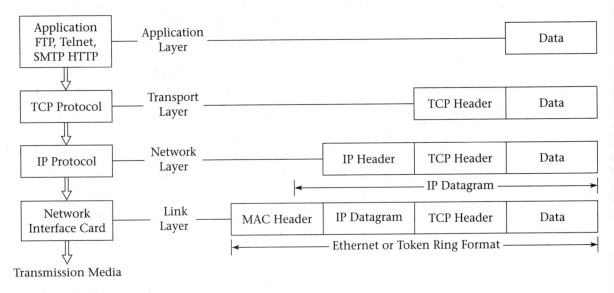

FIGURE 19.8 TCP operation

The following describe the function of each field in a TCP packet:

Sequence Number: The number label for each packet sent by the source

ACK Sequence Number: Acknowledges the next packet expected to be received from the source

Header Length: Identifies the length of the header in a 32-bit word

Flag bits: Six bits long and used for establishing a connection and activating a disconnection

URG	ACK	PSH	RST	SYN	FIN

Source Port 16 bits Identifies source application program such as Telnet = 23, FTP = 21, and SMTP = 25			Destination Port 16 bits Identifies which application program on the receiving side receives the data
Sequence Number 32 bits A number assigned to the packet by the source			
Acknowledgment Number 32 bits Acknowledges the next sequence number of the packet received from the source			
Header Length 4 bits Identifies number of 32-bit words in TCP header	Reserved 6 bits	Flag Bit 6 bits	Window Size 16 bits Size of the buffer source
TCP Checksum 16 bits Used for error detection in TCP header and data field		Urgent Pointer 16 bits Field is valid if URG bit in flag is set	
Data (if any)			

FIGURE 19.9 TCP packet format

URG: Urgent Pointer is set to '1' when that field contains urgent data

ACK: ACK bit is set to '1' to represent that the acknowledge number is valid

PSH: Set to '1' means the receiver should pass the data to an application as soon as possible

RST: Resets connection

SYN: Set to '1' when a node wants to establish a connection

FIN: Set to '1' means this is the last packet

Port Numbers   A **port number** is a logical channel in a communication system. The Transmission Control Protocol and User Datagram Protocol use port numbers to demultiplex messages to an application. Each application program has a unique port number associated with it. TCP/IP port numbers, which are between 1 and 1023, are well-known ports and these numbers are reserved for special applications by the Internet Authority. Table 19.2 shows a few commonly used port numbers.

19.5 Internet Level Protocols: IP and ARP

Internet level protocols consist of IP Protocol, Address Resolution Protocol (ARP), Reverse ARP, and Internet Control Message Protocol (ICMP).

Internet Protocol (IP)   The function of IP is packet delivery with unreliable and connectionless service. These Internet datagrams are also called IP datagrams. All TCP, UDP, ICMP, and ARP data are transmitted as IP datagrams. Figure 19.10 shows an IP datagram packet format.

TABLE 19.2 Commonly Used Port Numbers

Network Services	Port Number	Network Services	Port Number
tcpmux	1/tcp	netstat	15/tcp
echo	7/tcp	chargen	19/tcp
echo	7/udp	chargen	19/udp
discard	9/tcp	ftp-data	20/tcp
discard	9/udp	ftp	21/tcp
systat	11/tcp	telnet	23/tcp
daytime	13/tcp	smtp	25/tcp
daytime	13/udp		

0   31

IP Version 4 bits (current version is 4)	**Header Length** 4 bits Defines number of 32-bit words in the header	**Type of Service (TOS)** 8 bits Specifies how the datagram should be handled	**Total Length** 16 bits Specifies the length of IP datagram including the header in bytes
Identification 16 bits Used by destination to identify different datagram from one file	**Flags** 3 bits Currently uses the first 2 bits, DF and MF bits. DF = 1 Do not fragment; MF = 1 More fragment is coming		**Fragment Offset** 13 bits Indicates where in the original datagram this fragment belongs
Time to Live (TTL) 8 bits Specifies number of routers the datagram can pass	**Protocol** 8 bits Specifies the protocol which data belongs to, such as TCP, UDP, ICMP		**Header Checksum** 16 bits The 16-bit one's complement sum of the header
Source IP Address 32 bits IP address of sending machine			
Destination IP Address 32 bits IP address of receiving information			
Options if Any	**Padding**		
Data			

FIGURE 19.10 IP datagram packet format

The following describe the function of each field in an IP packet:

Version: Contains IP version number, which is 4 bits, and the current version number is 4 (IPv4).

Header Length: Represents the number of 32-bit words in the header. If there are no IP options and padding, header length is 20 bytes (5 words). IP is an unreliable service; there is no acknowledgment from the destination to the source. There is no physical connection between the source and destination. IP datagrams can arrive at the destination out of order.

Type of Services (TOS): TOS is eight bits. For most purposes, the values of all the bits in TOS are set to zero; meaning that in normal service the unused bits are always zeros.

0	1	2	3	4	5	6	7
Precedence			D	T	R	Unused	

Precedence indicates the importance of a datagram (0 normal; 1 next important).

TCP/IP Protocol ignores this field.

D, T and R identify the type of transport the datagram requests.

D is for Delay, T is for throughput, and R is for reliability.

D = 0 Normal delay

D = 1 Low delay

T = 0 Normal throughput

T = 1 High throughput

R = 1 High reliability

R = 0 Normal reliability

Total length: This field identifies the total length of the datagram (including the header) in bytes.

Identification: This is a number created by the sending node. This number is required when reassembling fragmented messages. The identification field is used by the destination to put together related datagrams.

Flag 3 Bits: The first bit is unused. The other two bits are: DF = Do Not Fragment, and MF = More Fragment. If DF is set to 1, the datagram cannot be fragmented. If the size of the data is more than the size of the MTU, the node cannot transfer data and an error message is generated.

0	1	2
Unused	DF	MF

Fragmented Offset Field: The offset field represents the offset of data in multiples of eight; therefore, the fragment size should be multiples of eight.

Example: 1,000 bytes are to be transferred over a network with an MTU of 256 bytes. Assume that the header of each datagram is 20 bytes. Find the number of datagrams if the following information is given.

- Identification: can be any number
- Total Length
- Frame Offset
- More Fragment

$256 - 20 = 236$ bytes

$8 \times 30 = 240$

$8 \times 29 = 232$

Each fragmented datum contains 232 bytes.

Identification	20	20	20	20	20
Total Length	232 + 20	232 + 20	252	252	92
Fragmented Offset	0	29	58	87	116
MF	1	1	1	1	0

Time To Live (TTL): This field indicates amount of time in seconds that a datagram may remain on the network before it is discarded. This value (32 or 64) is set by the sender and is decremented by one every time a router handles the datagram. If this field becomes zero, the datagram is thrown away.

Protocol Type: The number in the field identifies the High Level Protocol that generates this datagram or allows the destination IP to pass the datagram on to the required protocol. For example: ICMP is 1, UDP is 17, and TCP is 6.

Header Checksum: This field is the checksum of the header (not the data field). The checksum is the sum of the one's compliment of the 16-bit word of the header.

Sending Address: This is the IP address of the source.

Destination Address: This is the IP address of the destination.

Address Resolution Protocol

The interior gateway must have the physical address of the host connected to one of its local networks. The gateway must record both the IP address and its physical address. If the gateway does not have the physical address of the host, it will send an **Address Resolution Protocol (ARP)** packet to get the host's physical address. The host will respond through the RARP protocol. Figure 19.11 shows ARP packet format.

0   15

Hardware Type 16 bits	
Protocol Type 16 bits	
HLEN Hardware Address Length 8 bits	PLEN IP Address Length 8 bits
Operation Code 16 bits ARP Request = 1 RARP Request = 3	 ARP Response = 2 RARP Response = 4
Sender Hardware Address 48 bits	
Sender IP Address 32 bits	
Target Hardware Address 48 bits	
Target IP Address 32 bits	

FIGURE 19.11 ARP packet format

ARP is used by the interior gateway to access the physical address of the target station on the same network from the IP address of the network. Each station has a table located in cache called the ARP cache table.

If station A needs the physical address of B (as shown in Figure 19.12), station A will broadcast the APR message on the network, and any station matching its IP address with the IP address in the ARP packet will accept the ARP packet. Then it responds to the station A with RARP, which contains the physical address of station B.

FIGURE 19.12
ARP and RARP
architecture

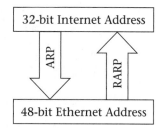

The following describe the function of each field in an ARP packet format:

Hardware Type: Hardware type identifies the type of hardware interface. The following are some of the hardware types:

Type	Description
1	Ethernet
2	IEEE 802.3
3	X.25
4	Token Ring

Protocol type: Protocol type identifies the type of protocol the sending device is using. For example: Protocol type 0800H is used for IP.

HELN: Hardware Address Length (HELN) in bytes (means $6 \times 8 = 48$ bits is the size of the hardware address).

PLEN: Length of IP address

Operation Code: Indicates whether the datagram is an ARP request or ARP response.

1 = ARP request

2 = ARP response

3 = RARP request

4 = RARP response

Maximum Transfer Unit

Maxium Transfer Unit (MTU) is the largest frame length it is possible to send on a given physical medium. There is a limit to the frame size. For example, 802.3's maximum frame size is 1500 bytes. If the datagram is larger than the MTU, the datagram is fragmented into several frames, each less than the MTU. Table 19.3 shows the MTU values for a few common network types.

TABLE 19.3   Network Types with MPU Values

Network	MTU (Bytes)
4 Mbps Token Ring	4464
16 Mbps Token Ring	17914
FDDI	4352
Ethernet	1500
IEEE 802.3	1492
X.25	576
Point-to-Point	296

19.6   IPv4 Addressing

An IP address is a 32-bit number which forms a unique address for each host connected to the Internet. No two hosts can have the same IP address. The assignment and maintenance of IP addressing is maintained by InterNIC.

An IP address is written in dotted decimal (Base$_{10}$) notation and is represented by four 8-bit binary numbers with the range from 0 to 255 ($4 \times 8 = 32$ bits).

Binary (Base$_2$)   00000000   to   11111111
Decimal (Base$_{10}$)   0   to   255

IP addresses are organized into the following five classes.

Class A Address   The Class A IP address is used for organizations with a large number of users connected to the Internet and a small number of networks.

Class A IP address format

	7 bits	24 bits
0	NET ID	HOST ID

The first most significant bit of a Class A IP address is zero.

The Network ID is 7 bits: 2^7 is 127.

The Host ID is 24 bits: $2^{24} = 16 \times 10^6$ nodes.

The range of a Network ID is from 0 to 127. The numbers 0 and 127 are reserved.

The range of Class A addresses is from 0.0.0.0 to 127.255.255.255 in dotted decimal.

Class B Address   A Class B address is used for medium-sized networks having more than 255 hosts.

The first two bits of a Class B address are 1 and 0.

The Network ID is 14 bits and the Host ID is 16 bits.

With a Class B address we can have 2^{14} (16,384) networks and each network can have 2^{16} (65,536) hosts or nodes.

The range of a Class B Network ID is from 128.0 to 191.255.

Class B IP address format

		14 bits	16 bits
1	0	NET ID	HOST ID

Class C Address   A Class C address is used for networks with a small number of hosts (those networks whose number of hosts does not exceed 255).

The first three bits of a Class C address are 1, 1, and 0.

Twenty-one bits are used for Network ID, and 8 bits are used for Host ID.

Class C IP address format

			21 bits	8 bits
1	1	0	NET ID	HOST ID

A Class C IP address can handle 2^{21} networks.

Each network can have 256 host IDs.

The range of a Class C Network ID is from 192.0.0 to 223.255.255.

The IP address of 192.0.2.1 was never assigned and is used for *test* purposes only.

Class D Address Class D address is reserved for multicasting. In multicasting, a packet is sent to a group of hosts.

Class D IP address format

1	1	1	0	MULTICAST GROUP ID (28 bits)

Class E Address Class E address is reserved for future use.

Loopback Address The last address of each class is used as a **loopback address** for testing; this loopback address is used on a computer to communicate with another process on the same computer.

The loopback addresses are:

Class A   127.0.0.1
Class B   191.255.0.0
Class C   223.255.255.0

Network Address The host portion of a network address is set to zero. For example: 129.49.0.0 is a **network address**, but it is not a node address. No node is assigned to 0.0.

Broadcast Address The host portion is set to all '1's in a broadcast address. A packet with a broadcast address is sent to every node in the network. For example: address 129.49.255.255 is a **broadcast address**.

19.7 Assigning IP Addresses

Most universities will have networks connected to the Internet as shown in Figure 19.13. The network administrator must contact InterNIC and obtain an IP address for the university. The InterNIC assigns a Class B address with a Network ID of 129.47 (the first two bytes of the IP address). The network administrator requires two bytes; from these two bytes he/she must decide how many bits are needed for each subnetwork. The answer is determined by how the network is growing and the future needs of the university. In this example we use one byte to represent our subnetwork address.

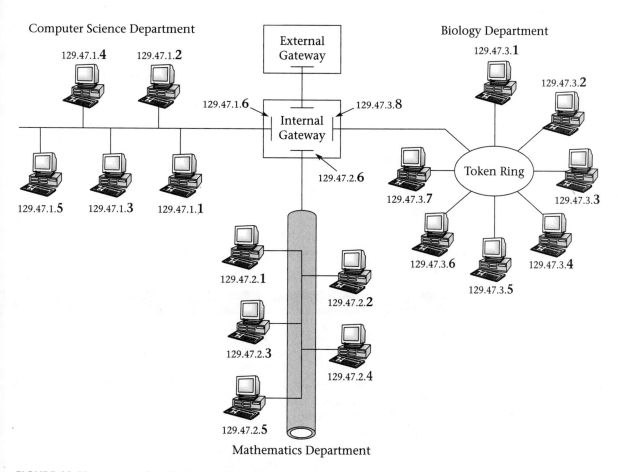

FIGURE 19.13 A network with three LANs and one gateway

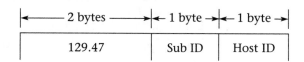

The subnetwork ID is eight bits, implying that we can have 256 networks in the university with 255 nodes each.

Subnetwork ID

01 Computer Science Department

02 Mathematics Department

03 Biology Department

04 to 255 reserved for future use

In Figure 19.13 you can see that the Computer Science Network ID is 129.49.1 and the host's IDs are assigned 1, 2, 3, 4, and 5 (shown in bold type).

The gateway uses the same Network Interface Card (NIC) that the network is connected to. If the network is Ethernet, the interface card inside the gateway will be an Ethernet card.

Address Mask   Address masks are used to define how many bits of an IP address are allocated for the host ID. It is used to separate the network address from a host ID. In the above example the least significant byte is used for a host ID as shown in Figure 19.14.

FIGURE 19.14
Address mask

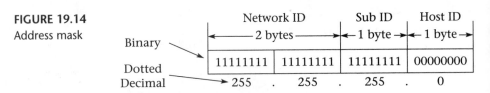

19.8   Point-to-Point Protocol

The **Point-to-Point Protocol (PPP)** provides a standard method for transmitting IP packets over serial point-to-point links such as modems. PPP has a number of advantages over SLIP (Serial Link Protocol). It is designed to operate over both asynchronous connections and bit-oriented synchronous systems. Since PPP is more advanced than SLIP, it can configure connections to a remote network dynamically and test that the link is usable.

PPP supports asynchronous links with eight bits of data and a bit-oriented link. Figure 19.15 shows the PPP packet format. The start and end flags are determined by 7E. The address field is always FF. The control byte is 03 and the Protocol field is two bytes. Table 19.4 shows some of the protocol types.

Flag 7E	Address FF	Control 03 for PPP	Protocol Type 0021 means information is IP datagram	Information	FCS	Flag 7E
1 byte	1 byte	1 byte	2 bytes	0–1500 bytes	2 bytes	1 byte

FIGURE 19.15   Point-to-point protocol packet format

TABLE 19.4   Some Protocol Types

Table Protocol Type	Information Field
0021	IP Datagram
C021	Link Control Information
8021	Network Control Information

The Link Control Information is used for establishing the link and negotiating options. The Network Control Information specifies network layers such as IP, Decnet, and so forth.

19.9 Demultiplexing Information

Figure 19.16 illustrates the general block diagram of Internet hardware and protocols. The packets (in the form of electrical signals) come to the Physical layer of the Network Interface Card. The Physical layer changes the signal to bits and passes it to the MAC sublayer. The MAC sublayer takes off its header (preamble, SFD, SA, and DA) and passes it to the Logical Link Control. The LLC checks the type field. If the type field is 0800H, the pack is an IPv4 datagram and is passed to IP.

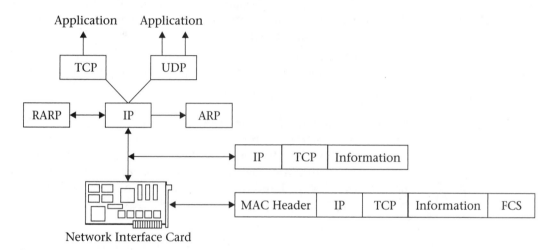

FIGURE 19.16 Demultiplexing of information

For IPv6 the type field would be 86DD H (hexadecimal). IP looks at the 8-bit protocol field (TCP = 6, UDP = 17, ICMP = 1, and IGMP = 6). IP will take off its header and pass the data to one of these protocols depending on the protocol number. Assuming the data is passed to TCP, TCP will look at the port number and pass it on to the Application layer.

19.10 TCP Connection and Disconnection

TCP Connection   TCP first establishes the connection with the destination before any information is transmitted, as shown in Figure 19.17. A connection is set up as follows:

FIGURE 19.17

Setting up a connection between two hosts

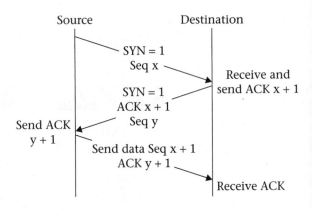

1. The source sends a packet to the destination by setting SYN = 1 in the TCP header and setting a sequence number in the TCP packet format (assuming the sequence number is x).

2. The destination will respond to the source by setting the SYN bit to 1, the acknowledge field to x + 1, and the sequence number to Y.

3. The source will start transmitting data.

TCP Disconnection When the source sends the last packet to the destination, the source sets FIN to 1 to inform the destination that this is the last packet. The destination acknowledges the last packet and sets the FIN to 1 to inform the source that destination does not have any packet to send. The source sends a packet with RS set to 1 and the destination responds to the source with a packet whose RS bit is set to 1. Figure 19.18 shows the disconnection process.

FIGURE 19.18

Disconnecting a TCP connection between two hosts

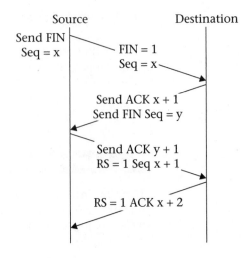

19.11 Internet Protocol Version 6 (IPv6)

Due to the growth of the Internet and the address limitations of IPv4, in 1995 the Internet Engineering Task Force (IETF) approved IPv6. These limitations which led to the version 6 upgrade are briefly summarized in the next paragraph.

The IPv4 address size is 32 bits and can connect up to $2^{32} = 4$ billion users to the Internet. The IPv4 address field is divided into two parts, the network address (Network ID) and host address (Host ID). Once a network number is assigned to an organization, the organization might not use all of the available host IDs in the host ID field; this means that all IPv4 addresses might not be used completely. Also, the number of networks connected to the exterior gateway increases rapidly and this causes the routing table to become large. When the routing table becomes large it takes more time to search through the table.

The IPv6 protocol reduces the size of the routing table in exterior gateways because IPv6 uses a hierarchical scheme to define an IP address. IPv6 has the following features:

- Expanded Addressing
- Simplified Header Format
- Support Extension
- Flow Labeling
- Authentication and Privacy

IPv6 Structure   IPv6 is divided into two parts: Basic Header and Extension Header. The first 40 bytes of the header are called the basic header as shown in Figure 19.19.

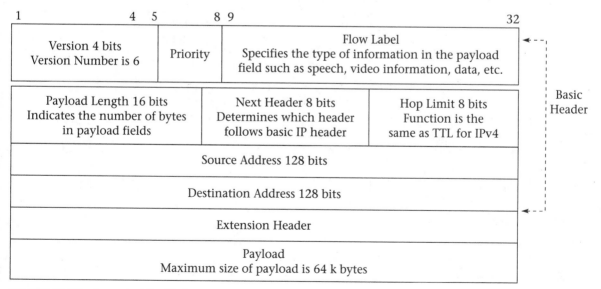

FIGURE 19.19   Format of the IPv6

IPv6 addresses are 128 bits in length. Addresses are assigned to individual interfaces on the nodes rather than to the nodes themselves. A single interface may have multiple addresses. The fields shown in Figure 19.19 are explained as follows:

Version: The version field is 4 bits, and it is 0110.

Priority: This field defines the priority of an IP datagram. It is used for congestion control. For example: E-mail has a priority of 2, and interactive traffic has a priority of 6.

Flow label: The flow label is 24 bits and is used for labeling packets that belong to particular traffic that the sender requests for special handling, such as real-time traffic.

Payload Length: 16 bits. Indicates the number of bytes in the payload fields.

Next Header: 8 bits. Determines which header follows the basic IP header.

Hop Limit: 8 bits. The function of this field is the same as TTL for IPv4.

IP Extension Header   Figure 19.20 shows an IP datagram with several extension headers and some of the header values. These values are also listed in Table 19.5.

In Figure 19.20 the first value of the Next Header is 0, which determines that the following header is Hop-by-Hop Options. The second value of Next Header is 43, which determines that the following header is Routing Information. The third value of Next Header is 6, which identifies the third header as TCP.

0			31
Version 4 bits	Priority 4 bits	Flow Label 24 bits	
Payload Length 16 bits		Next Header 0 8 bits	Hop Limit 8 bits
Source Address 128 bits			
Destination Address 128 bits			
Next Header 43 8 bits	Header Length 8 bits		
Hop-by-Hop Options			
Next Header 06	Header Length		
Routing Information			
TCP Header and Data			

FIGURE 19.20   IP datagram with several extension headers

TABLE 19.5   Header Values for Figure 19.20

Value	Header Function
0	Hop-by-Hop Option Header
4	IP
6	TCP
17	UDP
43	Routing Header
50	Encrypted Security Payload

IPv6 Address Architecture

IP Address Representation   The IPv6 address is 128 bits. It is divided into eight fields of 16 bits. Each field is represented in hexadecimal form. In general, the form of the IPv6 address is represented as:

Y: Y: Y: Y: Y: Y: Y: Y

Y is four digits in hexadecimal or 16 bits in binary. For example:

FE45:436A:12DF:4563:879E: 0008:0000:1232

In the preceding address, the leading zero of each field can be skipped. For example, the field containing 0008 can be written as 8 and the field containing all zeros can be represented by double colons. Only one double colon in each address is allowed. Therefore, the above address can be represented as:

FE45:436A:12DF:4563:879E: 8::1232

Addressing format   The specific type of address of IPv6 is indicated by the most significant bits. It is variable in length, and it is called the format prefix. Table 19.6 shows some of the format prefixes.

TABLE 19.6   Some Prefix Values for IPv6

Prefix Binary Value	Allocation
0000 0000	Reserved
0000 010	Reserved for IPX
011	Provider Unicast Address
1111 1111	Multicast Address

IPv6 Address Types   IPv6 supports the three types of addresses listed below.

1. **Unicast address:** This type of address is used for a single interface.
2. **Anycast address:** This type of address is used for more than one interface. A packet sent with an any cast address is delivered to only one of the interfaces—the one which is nearest to the source.
3. **Multicast address:** This type of address is used for a set of interfaces. A packet with a multicast address will transmit to all interfaces.

IPv6 Unicast Address   One of the most useful addresses is the provider-based unicast address. Its format is shown in Figure 19.21. This type of address is used for Internet provider services to an organization.

3 bits	5 bits	16 bits	16 bits	8 bits	32 bits	48 bits
Format Prefix	Registry ID	Provider ID	Subscriber Type	Subscriber ID	Subnetwork ID	Interface ID

FIGURE 19.21   Unicast address format

The following list is a breakdown of the address fields:

Format Prefix: The prefix for the provider-based unicast address is 011.

Registry ID: Identifies the registry that assigns the provider portion of the unicast address.

Provider ID: Identifies the specific provider who assigns to subscriber of the address.

Subscriber Type: Identifies the type of organization (education, corporation, etc.).

Subscriber ID: Identifies the specific subscriber from multiple subscribers.

Subnetwork ID: Identifies the network.

Interface ID: Identifies the interface. It is an IEEE 802 MAC address.

Loopback Address: The unicast address 0:0:0:0:0:0:0:1 is used by a node to send an IPv6 datagram to itself.

IPv6 Address with IPv4   Figure 19.22(a) shows the IPv6 address format that embeds IPv4 into the last 32 bits of IPv6. Another type of IPv6 unicast address used for nodes which do not support IPv6 addressing is shown in Figure 19.22(b).

FIGURE 19.22(a)
IPv4 embedded into IPv6

80 bits	16 bits	32 bits
0000....................0000	0000	IPv4 Address

FIGURE 19.22(b)
IPv4 embedded into IPv6 for a station
that does not support IPv6

80 bits	16 bits	32 bits
0000....................0000	FFFF	IPv4 Address

19.12 Internet II

vBNS is a cooperative project between MCI and the National Science Foundation (NSF) to provide a high bandwidth network for research in Internet technologies. The National Science Foundation and MCI Telecommunication Corp jointly provide **very High Performance Backbone Network Service (vBNS)** to five NSF supercomputer Centers (SCCs). These five supercomputer centers are:

- Cornell Theory Center (CTC)
- National Center for Atmospheric Research (NCAR)
- National Center for Supercomputing Applications (NCSA)
- Pittsburgh Supercomputing Center (PSC)
- San Diego Supercomputing Center (SDSC)

vBNS provides service to the above centers and other institutions for research and development of new technology for the Internet. These institutions are connected to vBNS by high capacity interconnection points called Gigabit Capacity Point-of-Presence or **GigaPOP**. vBNS uses advanced technology such as ATM and SONET for transmitting video, voice, and data. It operates at a speed of 622 Mbps using OC-12, as shown in Figure 19.23.

Summary

- The **Internet** is a collection of networks connected together through gateways and routers.
- The Internet uses a set of protocols for communications called **Transmission Control Protocol** and **Internet Protocol (TCP/IP)**.
- The applications of the Internet are: E-mail, Telnet, File Transfer Protocol (FTP), and Hypertext Transfer Protocol (HTTP).
- An **e-mail** address is made up of the user name and mail server such as *Elahi@scsu.ctstateu.edu*.
- **Telnet** is one of the most useful Internet applications. It enables one computer to remote login to another computer.
- File Transfer Protocol (FTP) is a protocol used for transferring a file from/to a remote computer without logging in to the remote computer.
- Hypertext Transfer Protocol (HTTP) is an advanced file retrieval protocol that can access distributed documents on the web.

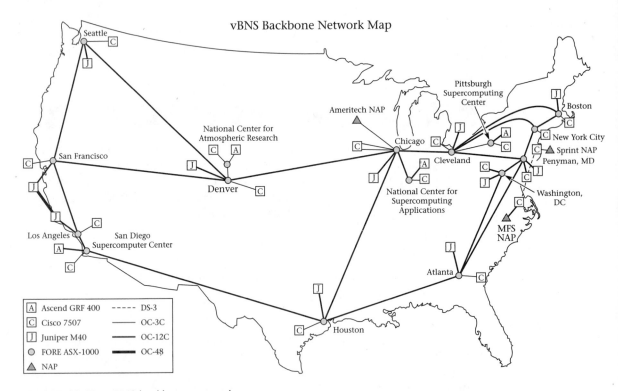

FIGURE 19.23 vBNS backbone network map

- Transmission Control Protocol (TCP) provides reliable service between hosts.
- Internet Protocol (IP) provides packet delivery between the hosts.
- Internet Address Version 4 (IPv4) is 32 bits and each byte is represented by a decimal number and divided into classes A, B, C, D and E.
- Class A IP Address range is from 0.0.0.0 to 127.255.255; the network IDs of 0 and 127 are reserved.
- The range of Class B IP Address is from 128.0.0.0 to 191.255.255.255.
- The range of Class C IP Address is from 192.0.0.0 to 223.255.255.
- The Class D IP Address is reserved for multicasting and Class E IP Address is reserved for future use.
- **Domain Name System (DNS)** is a distributed database containing domain name and corresponding IP address. Top domain names are: EDU, GOV, COM, MIL, ORG, and INT.
- **User Datagram Protocol (UDP)** provides unreliable service between hosts Applications of UDP are Trivial File Transfer Protocol (TFTP) and Remote Procedure Call (RPC).

- **Point-to-Point Protocol (PPP)** is used for transferring IP packets over serial link such as modems.
- The TCP header is 20 bytes and the IP header is 20 bytes.
- **IPv6** or **IP next Generation** address size is 128 bits.
- **Internet II** is used for research and for developing new technology for the Internet.
- Internet II uses **very High performance Backbone Network Service (vBNS)** for its backbone.
- vBNS operates at the speed of 622 Mbps using OC-12.

Key Terms

Address Mask
Address Resolution Protocol (ARP)
Advanced Research Project Agency (ARPA)
Anonymous FTP
Anycast Address
ARPANET
Broadcast Address
Domain Name
Internet Protocol (IP)
Internet Protocol Version 6 (IPv6)
Loopback Address
Maximum Transfer Unit (MTU)

Multicast Address
Network Address
NSFNET
Point-to-Point Protocol (PPP)
Port Number
Remote Login
Telnet
Transmission Control Protocol (TCP)
Unicast Address
User Datagram Protocol (UDP)
very High Performance Backbone Network Service (vBNS)

Review Questions

- **Multiple Choice Questions**

    1. The Internet is a collection of _____.
       - a. servers
       - b. applications
       - c. networks
       - d. routers

    2. The Internet uses the _____ protocol.
       - a. X.25
       - b. NWLink
       - c. TCP/IP
       - d. Windows NT

3. Which of these is not an application of the Internet? _____
   a. E-mail   c. WWW
   b. FTP   d. Antivirus software

4. _____ provides packet delivery between networks.
   a. IP   c. X.25
   b. TCP   d. ARP

5. The Transport layer of TCP/IP consists of UDP and _____.
   a. TCP   c. a and b
   b. IP   d. ICMP

6. _____ performs reliable delivery of data.
   a. IP   c. TCP
   b. UDP   d. RARP

7. Which organization ratifies standards for the Internet? _____
   a. IEEE   c. IETF
   b. ITU   d. EIA

8. What type of switching is used in the Internet? _____
   a. Virtual circuit   c. Circuit switching
   b. Packet switching   d. Message switching

9. Which protocol is used for unreliable communication? _____
   a. TCP   c. IP
   b. UDP   d. ARP

10. Telnet uses which of the following protocols for remote login? _____
    a. UDP   c. IP
    b. TCP   d. FTP

11. Which protocol is used by a modem to connect a computer to the Internet? _____
    a. TCP/IP   c. PPP
    b. UDP   d. ARP

12. Telnet enables a user to _____.
    a. transfer a file   c. remote login
    b. send E-mail   d. transfer mail

13. How many bits is an IPv4 address? _____
    a. 24 bits   c. 48 bits
    b. 32 bits   d. 128 bits

14. What is the application of a loopback address? _____
    a. Reserved by Internet authority   c. Used for broadcast address
    b. Used for testing   d. Used for unicast

15. Which of the following addresses is a Class C broadcast address? _____
    a. 191.205.205.255   c. 191.205.205.00
    b. 191.205.205.205   d. 191.205.255.255

16. What protocol is used for the World Wide Web? _____
    a. TCP/IP   c. UDP
    b. HTTP   d. ARP

17. TCP is used for _____.
    a. reliable communication
    b. unreliable communication
    c. connection-oriented communication
    d. none of the above

18. What is the function of the source and destination port in a TCP header? _____
    a. It is used to identify the source and destination host on the network.
    b. It is used to identify the application source protocol and application of destination protocol.
    c. It is used to identify source protocol and destination protocol.
    d. none of the above

19. What is the function of the Window Size field in a TCP header? _____
    a. The source reports its buffer size to the destination.
    b. The source reports the number of packets received from the destination.
    c. The source reports the size of its cache memory to the destination.
    d. The source reports an error in the packet.

20. What is the function of Time To Live (TTL) in a TCP header? _____
    a. It holds the time of the day.
    b. It defines the number of routers a datagram can pass.
    c. It defines the transmission time of a datagram between source and destination.
    d. It defines the number of words in a packet.

21. Which protocol is used to set up a connection between source and destination? _____
    a. UDP
    b. TCP
    c. IP
    d. ARP

22. How many bits is IPv6? _____
    a. 32 bits
    b. 48 bits
    c. 64 bits
    d. 128 bits

23. The Internet routes a datagram from one gateway to another base on the datagram by using the _____.
    a. IP address
    b. MAC address
    c. port address
    d. none of the above

24. IPv6 is represented by _____.
    a. decimal
    b. hexadecimal
    c. octal
    d. binary

25. What is the function of an IP subnet mask? _____
    a. An IP subnet mask represents bits in the network portion of an IP address.
    b. An IP subnet represents the bits in the host portion of an IP address.
    c. a and b
    d. none of the above

● **Short Answer Questions**

1. What is the markup language used to create files for the WWW?

2. What is the protocol used by the WWW to transfer a file?

3. What is the protocol used to access the Internet by a modem?

4. List the DNS indicators.

5. List the protocols in the Transport level.

6. List the protocols in the IP layer.

7. What is the size of an IPv4 address?

8. What is the size of an IP header?

9. Explain the function of IP.

10. Explain the function of TCP.

11. What is the size of a TCP header?

12. What is the function of the TTL field in an IP header?

13. What is the size of a UDP header?

14. List IP address classes.

15. If an organization uses the third byte of an IP address for a subnet, how many subnets can be assigned to the Class B address?

16. What is the name of the organization you apply to, to get an IP address?

17. What is the function of ARP?

18. Show the IP frame format.

19. What is the current version of IP?

20. What is function of the TTL field in an IP packet?

21. What is the size of an IPv6 address?

22. Convert the following IP address from hexadecimal to dotted decimal representation, and find the class type of each IP address.
    a. 46EF3A94
    b. 23446FEC

23. List Internet applications.

24. The following e-mail address is given. Identify user name and mail server address:

    Elahi@Xycorp.com

25. List two application protocols for UDP.

26. List three application protocols for TCP.

27. TCP/IP was developed by _____.

28. Show the TCP/IP Reference Model.

29. What is the function of ICMP?

30. What is DNS? Describe its function.

31. Explain the function of Telnet.

32. Show the TCP/IP Reference Model and list the protocols for each level.

33. What is a port number?

34. What is MTU?

35. Identify the class of the following IP addresses:
    a. 129.234.12.08
    b. 252.243.56.89
    c. 92.92.92.92

36. What is the data rate of vBNS?

37. Assume the Internet Network Information Center assigns you a Class B IP address 172.200.0.0.
    a. How many bits do you use from host ID for 128 subnet ID?
    b. How many host IDs can be generated for each subnet ID?
    c. What is the subnet mask ID?

38. There are two computers, A and B, with IP addresses of 174.20.45.37 and 174. 20.67.45, respectively. If these two computers both have a subnet mask ID of 255.255.0.0, can you determine if these two computers are located in the same network?

chapter 20

Asynchronous Transfer Mode

OBJECTIVES

After completing this chapter, you should be able to:

- List the components of the Asynchronous Transfer Mode (ATM)
- Describe the advantages of ATM
- Describe the applications of ATM
- Show ATM cell format
- Explain ATM connection types and connection identifiers
- Explain ATM switch characteristics
- Discuss ATM switch architecture
- Explain blocking in an ATM switch
- Show the ATM User Network Interface (UNI) cell header and understand the function of each field
- Show the ATM Network-to-Network Interface (NNI) cell header and understand the function of each field
- Show ATM end user and ATM switch protocol
- Explain the functions of the ATM Adaptation Layer, ATM Layer, and Physical Layer

- Describe the application of each ATM Adaptation layer
- Show AAL1, AAL2, AAL3/4, and AAL5 cell formats

Asynchronous Transfer Mode (ATM) is the next generation of networking technology to be used on the Information Superhighway. A much wider array of information can be transmitted using this technology, for example: voice, data, images, CAT scans, MRI images, and video conferencing. Both private and public networks are supported and voice, video, and data are transported on a common circuit. ATM delivers bandwidth on demand, is not dependent on applications, and works at a data rate from 1.5 Mbps to 2 Gbps. All types of networking, from LANs to WANs and from backbone to desktop, can integrate ATM technology. In addition, ATM is a transfer protocol for B-ISDN. Figure 20.1 illustrates a typical ATM network consisting of switches and end users. ATM switches offer two types of interfaces: switch-to-switch interface (or Network-to-Network Interface, called NNI) and Switch-to-User Interface, called UNI.

FIGURE 20.1
ATM network
interface

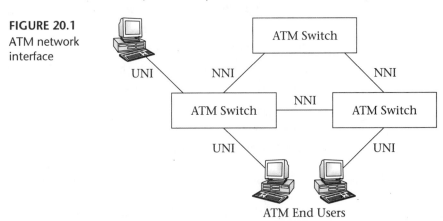

ATM End Users

20.1 ATM Network Components and Characteristics

The ATM network components are an ATM Network Interface Card, an ATM switch to relay each cell to its destination, and software.

The most significant advantages of ATM are as follows:

- Bandwidth on demand to users
- Connection-oriented service
- ATM switches use statistical multiplexing
- Higher quality of service
- Varied transmission media such as use of optical cable or twisted-pair wire
- Wider array of information can be handled
- Works with current LAN and WAN technologies and supports current protocols such as TCP/IP

The characteristics of Asynchronous Transfer Mode are listed below.

Cell Switching: A cell carries information about routing; the switch reads the information and selects the route.

Connection-Oriented Transmission: A connection must be established between two stations before data can be transferred between them.

Cell-Size: Fixed at 53 bytes, with a 5-byte header and a 48-byte payload, as shown in Figure 20.2.

FIGURE 20.2
ATM cell

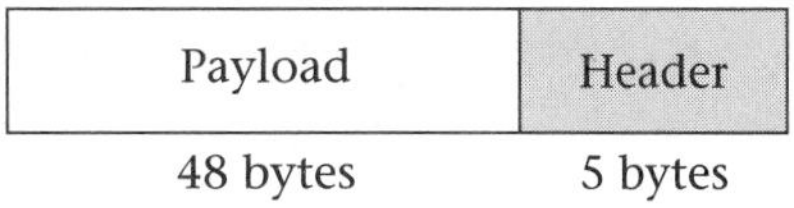

Payload	Header
48 bytes	5 bytes

Delay: Between transmissions is very small.

Switching: Occurs at very high speeds.

20.2 ATM Forum

The **ATM forum** started in 1991 with four computer and networking vendors. Today it has over 1,000 members. The forum writes specifications and definitions for ATM technology, and these specifications are submitted to the ITU for approval. The ITU-T standard organization works closely with the ATM forum.

20.3 Types of ATM Connection

ATM offers two types of connections:

1. **Permanent Virtual Connection (PVC):** PVC is set up and taken down manually by a network manager. A set of network switches between the ATM source and destination are programmed with predefined values for VCI/VPI. Transmission is more reliable with this type of connection.

2. **Switched Virtual Circuit (SVC):** SVC is a connection that is set up automatically by a signaling protocol. SVC is more widely used because it does not require manual setup, but it is not as reliable.

Connection Identifiers   There are two **connection identifiers** in an ATM cell header: the **Virtual Path Identifier (VPI)** and the **Virtual Channel Identifier (VCI)**. These are used for routing and identification of cells. VPI and VCI do not represent destination addresses, rather they represent a connection that leads to the intended destination. One VPI may contain several VCIs, as shown in Figure 20.3.

FIGURE 20.3
ATM connection identifiers

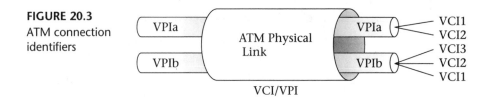

Figure 20.4 shows a representation of VPI and VCI. VPI is the railroad number and VCI is the wagon number. Each rail can transport several wagons. Each wagon has a unique number that is represented by its VCI. When a cell enters the switch, the switch can assign a new VCI number to the cell.

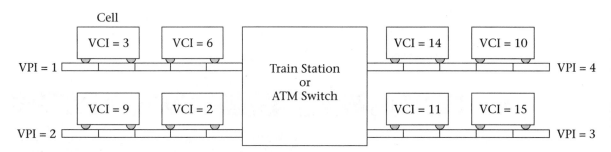

FIGURE 20.4   Representation of VPI and VCI in real world

20.4   ATM Switch Operation

An ATM switch can process cells at an extremely high rate of speed. The ATM switch operates by performing the following functions:

1. A cell is received on an input port. Its header VPI/VCI is examined to determine the output port to which the cell should be forwarded.
2. VPI/VCI fields are modified to a new value for the output port.
3. The Header Error Control (HEC) is used for error detection and correction in the header field of each cell. If the HEC cannot correct the error, the ATM switch will discard the cell.
4. The switch has a control unit which can modify the routing table.
5. The switch must support cell switching at a rate of at least one million cells per second.

Figure 20.5 shows an ATM switch with 3 ports. The cells enter the switch from port 1, and go out from ports 2 and 3 according to the routing table. The ATM switch has a routing table as defined by the ATM signaling protocol. Assume that

Table 20.1 represents the routing table for Figure 20.5. Any cell from port 1 with VPI/VCI = 1/26 entering the switch will go out through port 2 with VPI/VCI = 2/45. Any cell entering from port 1 with VPI/VCI = 1/45 will go out through port 3 with VPI/VCI = 3/39.

FIGURE 20.5
ATM switch with three ports

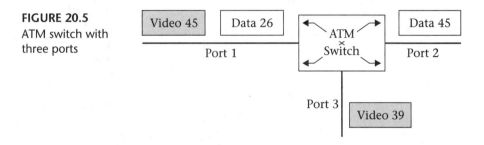

TABLE 20.1   Routing Table for Figure 20.5

Input		Output	
Port	VPI/VCI	Port	VPI/VCI
1	1/26	2	2/45
1	1/45	3	3/39

ATM Switch Characteristics   ATM switches should have the following characteristics:

- Have enough buffer capacity to store incoming cells and the routing table
- Be fast enough to transfer incoming cells to the output ports
- Have from 16 to 32 input and output ports
- Support all AAL types
- Support permanent virtual circuit (PVC) and switched virtual circuit (SVC) connections
- Support point-to-multipoint connections
- Act as support for congestion control (able to overcome congestion)

20.5   ATM Switch Architecture

The most important component of an ATM network is the switch, which must capable of processing billions of cells per second. Figure 20.6 shows the general architecture of an ATM switch using Statistical Packet Multiplexing (SPM). SPM dynamically allocates bandwidth to the active input channels. The function of the control processor is to control the input/output buffers and update the routing table of the switch. There are several different architectures used for ATM switches, such as the Delta switch matrix and the Banyan switch matrix.

FIGURE 20.6
General architecture
of an ATM switch

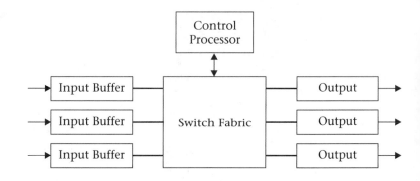

ATM Switch Blocking

An ATM switch will be involved with two types of blocking or traffic congestion. These are fabric blocking and head of the line blocking.

Fabric blocking   occurs when the fabric capacity of a switch is less than the sum of its input data rate. In this case the switch must drop some of the cells. Some ATM switches are limited to 16 or 32 OC-3 input ports.

Head of the line blocking   occurs when an output port is congested and a cell is waiting in the input port. The switch must drop some of the cells in the output port. Some switches randomly discard the cells, and all stations must retransmit all the cells. Also, some switches have intelligent systems that drop cells belonging to one source.

20.6   ATM Connection Setup Through ATM Signaling

ATM signaling is initiated by an end user who wants to set up a connection through an ATM network to a destination. The signaling packet is sent through the network from switch to switch with VPI = 0 and VCI = 5, and connection identifications are set up for each switch until the packet reaches its destination point. Figure 20.7 shows the first ATM user making a connection with an ATM switch or network. The ATM network then makes a connection with the ATM end user.

20.7   ATM Cell Format

ATM uses VLSI technology to segment data to the cell at high speeds. Each cell consists of 53 bytes, in which there are a 5-byte header and a 48-byte payload (as shown in Figure 20.8).

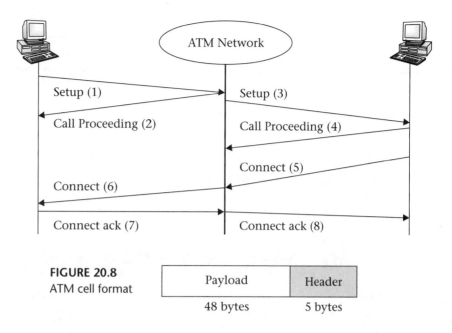

FIGURE 20.7
Connection setup through ATM signaling

FIGURE 20.8
ATM cell format

ATM defines two header formats: one for the User Network Interface (UNI) cell header and another for the Network-to-Network Interface (NNI) cell header.

UNI Cell Header   The User Network Interface (UNI) cell header defines communication between ATM end stations (workstation or router) and ATM switches. Figure 20.9 illustrates the UNI cell header format.

The following are the fields shown in Figures 20.9 and 20.10:

GFC: 4-bit Generic Flow Control can be used to provide local functions, such as identifying multiple stations sharing a single ATM interface. This is currently not used.

VPI: 8-bit Virtual Path Identifier used with VCI to identify the next destination of a cell as it passes through a series of ATM switches.

VCI: 16-bit Virtual Channel Identifier used with VPI to identify the next destination of a cell as it passes through a series of ATM switches.

PT: 3-bit Payload Type. The first bit indicates whether the payload is data or control data. The second bit indicates congestion, and the third bit indicates whether the cell is the last cell in the series that represents the AAL5 frame.

CLP: 1-bit Congestion Loss Priority, which indicates whether a cell should be discarded if it encounters extreme congestion. This bit is used for quality of service (QoS).

HEC: 8-bit Header Error Control. The checksum is used to calculate only the header. This is an 8-bit CRC that can detect all single errors and certain multiple bit errors. It can correct single bit errors.

FIGURE 20.9
UNI cell header
format

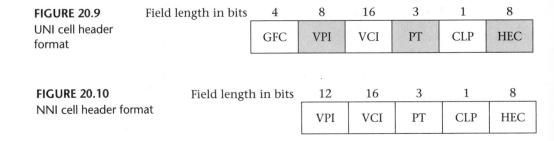

Field length in bits

4	8	16	3	1	8
GFC	VPI	VCI	PT	CLP	HEC

FIGURE 20.10
NNI cell header format

Field length in bits

12	16	3	1	8
VPI	VCI	PT	CLP	HEC

NNI Cell Header The **Network-to-Network Interface (NNI)** defines communication between ATM switches. The format of a NNI cell header is shown in Figure 20.10. The VPI field is 12 bits, allowing ATM switches to assign a larger value to the VPI.

20.8 ATM Protocol

Figure 20.11 shows an ATM end point operational model and an ATM switch operational model. The ATM end points consist of three layers: the ATM Adaptation layer (AAL), the ATM layer, and the Physical layer. An ATM switch consists of an ATM layer and a Physical layer. The ATM Adaptation layer, ATM layer, and Physical layer are divided into sublayers, as shown in Table 20.2.

FIGURE 20.11
ATM end points
connected to an
ATM switch

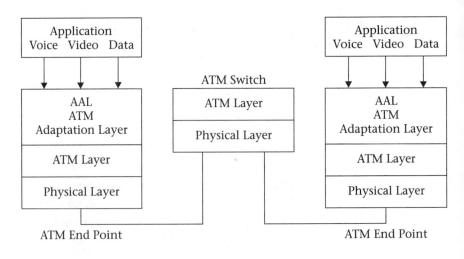

Physical Layer Functions As shown in Table 20.2, the Physical layer is divided into two sublayers, the Physical Medium Dependent (PMD) layer and the Transmission Convergence (TC) layer.

Physical Medium Dependent (PMD) Functions The PMD sublayer provides bit transmission, coding, electrical and optical conversion, and bit timing (the generation of signals suitable for transmission media).

TABLE 20.2 ATM Sublayers

Layer	Sublayer	Functions
ATM Adaptation Layer	Convergence Sublayer (CS)	Provides service to the lower and upper layers dependent on the type of AAL.
	Segmentation and Reassembly (SAR)	Segmentation of information to 48 bytes
ATM Layer		Cell Header Generation and Extraction Cell Demultiplexing and Multiplexing VPI/VCI translation
Physical Layer	Transmission Convergence (TC)	Cell Delineation, HEC Generation and Cell Verification. Add idle cell if necessary.
	Physical Medium Dependent (PMD)	Bit Timing, Framing, and Physical Transport

Many LAN technologies, such as Ethernet and Token Ring, specify certain transmission media. Optical cable, coaxial cable, and twisted-pair cable can all be used as ATM transmission media.

The ATM forum has defined different types of transmission media, such as SONET, which uses Multimode Fiber (MMF) cable and Single Mode Fiber (SMF) cable. Multimode fiber cable uses the same 4B/5B encoding as FDDI and twisted-pair cable. Table 20.3 gives examples of physical transmission media for ATM.

TABLE 20.3 Transmission Media for ATM Network

Layer	Rates (Mbps)	Media Type
OC-48	2488	SMF
OC-24	1244	SMF
OC-12	622.08	SMF
STS-12C	662.08	UTP-5
STS-3	155.52	MMF and UTP-5
STS-1	51.84	UTP-3
DS-3	45	Coaxial Cable

SMF = Single Mode Fiber   OC-48 = Optical Carrier
MMF = Multimode Fiber   STS-1 = Synchronous Transport Signal-1
UTP-3 = Unshielded Twisted-Pair Cat-3   DS-3 = Digital Signal

Transmission Convergence Sublayer (TC) The functions of the TC sublayer are as follows:

- Extracting the cell from the physical layer (extracting cells from the SONET envelope).
- Cell delineation—scrambling the cell before transmission and descrambling the cell after transmission.
- The TC receives a cell from the Physical layer, calculates the HEC header, and then compares it with the cell's HEC header. The TC uses the results of the comparison for error correction in the cell header. If the error cannot be corrected, the cell will be discarded.
- The Transmission Convergence sublayer receives the cell from the ATM layer and generates HEC; it then adds HEC to the cell.

ATM Layer The ATM layer performs the following functions:

- **Cell Header Extraction and Generation:** Add ATM cell header (except HEC) and remove cell header of incoming cells in an end station.
- **VPI/VCI translation:** Done in the ATM switch. The VPI/VCI value of an incoming cell is translated to a new VPI/VCI value for the outgoing cell.
- **Generic Flow Control:** Done on the user side (User-to-Network Interface) to determine the destination of the receiving cell.
- **Multiplexes and Demultiplexes cells:** Can be done from several different connections at the ATM switch.

ATM Adaptation Layer (AAL) The **ATM Adaptation layer** converts the large Service Data Unit (SDU) data packet of the upper layer to 48 bytes for the ATM cell payload. The AAL is designed so that the ATM can become more flexible, and more able to handle all types of traffic. The ATM Adaptation layer is divided into two sublayers: the convergence sublayer (CS) and the Segmentation and Reassembly sublayer (SAR).

After a connection is set up by the ATM signaling protocol, the convergence sublayer accepts higher layer traffic for transmission. The SAR segments each packet received from the CS into smaller units and adds a header or trailer, depending on the type of AAL, to form 48 bytes of payload.

ATM can be used for various applications. Therefore, different types of AALs are needed to provide service to upper layer applications. The AALs are divided into four classes of traffic. The ATM layer offers four types of AAL protocol. Each AAL protocol is used for a specific application, from class A through class D, as shown in Table 20.4.

TABLE 20.4 AAL Types and Their Application

Application Type	Class A	Class B	Class C	Class D
AAL Protocol	AAL1	AAL2	AAL3/4 and AAL5	
Timing Relation	Required	Required	Not Required	
Bit Rate	Constant Bit Rate	Variable Bit Rate		
Connection Type	Connection Oriented	Connection Oriented		Connection-less
Application	Audio	Compressed Video	Data	

The following list includes a brief description of each Class:

Class A: Constant Bit Rate (CBR). Connection-oriented, with a timing relation between source and destination required. Applied to uncompressed voice and video.

Class B: Variable Bit Rate (VBR). Connection-oriented, with a timing relation between source and destination required. Applied to compressed video and audio.

Class C: Variable Bit Rate (VBR). Connection-oriented, with a timing relation between source and destination not required. Applied to data.

Class D: Variable Bit Rate (VBR). Connectionless, with a timing relation between source and destination not required. Applied to connectionless data transfer.

20.9 Types of Adaptation Layers

ATM networks offer four different adaptation layers, in order to handle different types of traffic. The types of ATM Adaptation layers are AAL1, AAL2, AAL3/4, and AAL5.

Adaptation Layer Type 1

ATM Adaptation Layer Type 1 (AAL1) is designed to carry Class A traffic with constant bit rates (CBR), such as circuit emulation (which provides the same kind of services as a traditional leased-line or a time-division multiplexer), and is used for uncompressed video and telephone traffic applications. It is important not to misorder cells for voice communication; therefore, a sequence number is added in the SAR sublayer. Figure 20.12 shows the AAL1 cell format.

The following list describes the functions of the SAR header in Figure 20.12.

SN: The Sequence Number is 4 bits. The SC field is used for cell sequence numbers to detect any missing cell or misordered cell. The C bit is used for timing information.

SNP: Sequence Number Protection Field. The FCS bit is used for error detection on an SN field using a CRC polynomial $X^3 + X^2 + 1$. The P bit is used as a parity bit for the FCS field.

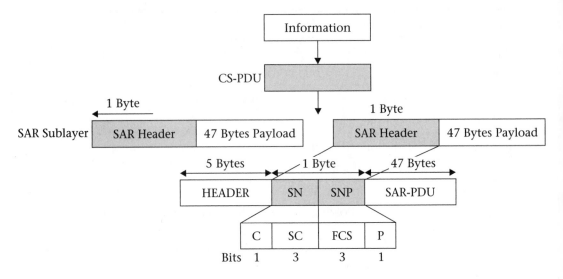

FIGURE 20.12 AAL1 cell format

Adaptation Layer Type 2 ATM Adaptation Layer Type 2 (AAL2) is used for variable bit rate applications such as compressed voice and audio.

Figure 20.13 shows the AAL2 cell format. Information from the Application layer is passed to the SAR sublayer. The SAR sublayer segments the receiving information into 45 bytes of payload and adds a 1-byte header and a 2-byte trailer to each payload. Then SAR passes them to the ATM layer.

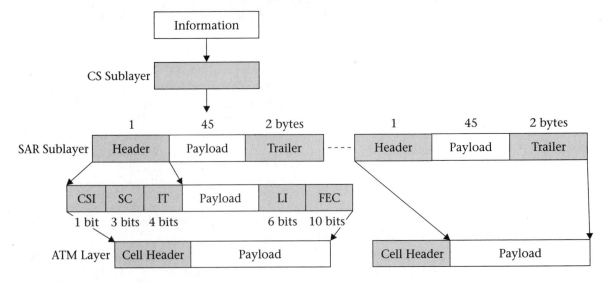

FIGURE 20.13 AAL2 protocol data unit

The following describe the function of each field of the SAR header:

CSI: Convergence sublayer identification is not defined yet.

SC: Sequence Count is used to detect lost cells.

IT: Information Type is used to indicate the position of payload, relative to the entire message. The IT indicates the beginning of the message, continuation of the message, and end of the message.

LI: Length Indicator is used to indicate the length of the message.

FEC: Forward Error Correction is used to correct some errors.

Adaptation Layer Type 3/4

ATM Adaptation Layers Types 3 and 4 (AAL3/4) started out as two separate Adaptation layers (AAL3 and AAL4). The specification of two AALs merged and the new type is called AAL3/4.

AAL3/4 is designed to take variable length frames up to 64 k bytes and segment them into cells. ITU recommends the AAL3/4 for data that is sensitive to loss but not to delay. The AAL3/4 can be used in connection-oriented transmission. Figure 20.14 shows the AAL3/4 format. The convergence sublayer accepts information from the upper layer and adds four bytes to the CS header and four bytes to the CS trailer (this is called CS-PDU). The CS transfers CS-PDU to the SAR sublayer and the SAR sublayer segments the CS-PDU into a 44-byte payload. The SAR adds a 2-byte SAR header and a 2-byte SAR trailer to each payload.

The function of each field in CS-DPU is listed below:

Type Field: Set to zero.

Btag and **Etag:** The two fields must have the same number to indicate the start and end of CS-PDU to the receiver. The transmitter changes the Etag and Btag for each successive CS-PDU.

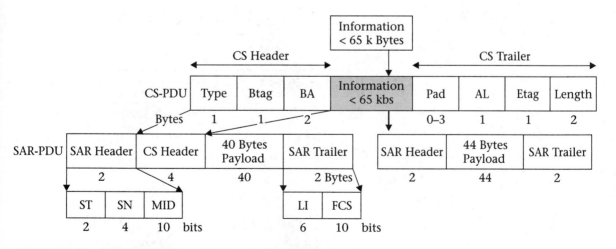

FIGURE 20.14 AAL3/4 CS sublayer PDU and SAR sublayer PDU

BA: Buffer Allocation, used by the destination to allocate buffer space for incoming cells.

PAD: Field is 4 bytes and is used to make the total number of bytes in CS-PDU multiples of four bytes.

Length: Indicates the size of a complete CS-PDU.

SAR: Sublayer segments the CS-PDU to 44 bytes and adds its header and trailer to each segmented CS-PDU (as shown in Figure 20.14) to form the 48 bytes of an ATM cell.

ST: Segment Type indicates if the cell is first, last, or a continuation.

AL: Alignment Byte is a dummy byte to make the trailer four bytes.

SN: Sequence number for detecting missing cells.

MID: Multiplexer Identifier.

LI: Indicates the number of useful bytes in the SAR-PDU.

FCS: Frame Check Sequence is used for error detection.

The above 48 bytes are passed to the ATM layer, which adds a 5-byte header to it.

Adaptation Layer Type 5

ATM Adaptation Layer Type 5 (AAL5) is an efficient way to transfer information in ATM, because all control information is in the last cell. The AAL5 format can be used for LAN emulation and can handle very large Ethernet and Token Ring frame formats. AAL5 is used for variable lengths of information up to 65 kbs. These 65 kbs are segmented into a series of cells, each with the same header. The last cell arrives at the destination carrying all the control information needed to handle the packet. Figure 20.15 shows the AAL5 cell format. AAL5 provides services similar to those AAL3/4 provides, but uses fewer control fields.

FIGURE 20.15
AAL5 cell format

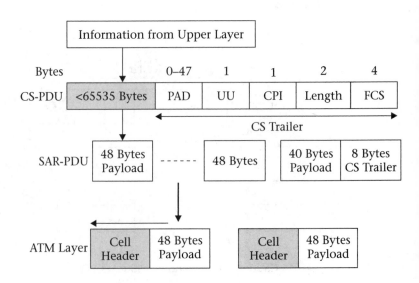

Information from upper layers passes to the CS. The CS adds a trailer to the information to generate a CS-PDU. The CS then passes the CS-PDU to the SAR for segmentation of up to 48 bytes of cells. The segmented cells are passed to the ATM layer. The ATM layer adds a 5-byte header to each cell and passes the cells to the Physical layer for transmission.

The following describe the function of each field of the convergence sublayer (CS) trailer.

UU: AAL layer User-to-User Identification

CPI: Common Port Identification (it is not defined)

Length: Indicates number of bytes in payload or data field

FCS: Used for error detection in CSC-PDU

20.10 Comparing ATM with Gigabit Ethernet

ATM can be used for both LANs and WANs, though it is not popular for LANs because it is expensive and complex. Most LAN designers prefer Gigabit Ethernet to ATM. However, unlike Gigabit Ethernet, ATM supports a constant bit rate for video and image transmission, provides real-time communication, and offers quality of service (QoS). Adapting the QoS protocol (IEEE 802.1Q) in the upper layer of Gigabit Ethernet enables it to prioritize packets, thereby assigning high priority to packets used in real-time communication and low priority to packets that are not time-sensitive.

Gigabit Ethernet has grown so rapidly that it has become one of the most popular LANs; whereas ATM is more widely used for WANs.

Summary

- **Asynchronous Transfer Mode (ATM)** is the next generation of networking technology which can handle all types of information such as data, voice, video, and images.
- An ATM network consists of an ATM switch and an ATM end user.
- The advantages of ATM networks are variable bandwidth, Quality of Service (QoS), ability to handle all types of traffic, high bandwidth, and small cell size.
- An ATM cell is made up of a 48-byte payload and 5-byte header.
- ATM offers two types of connections: **Permanent Virtual Connection (PVC)** and **Switched Virtual Connection (SVC)**.
- The connection identifiers in an ATM cell are **Virtual Path Identifier (VPI)** and **Virtual Channel Identifier (VCI)**.
- Permanent Virtual Connection (PVC) involves manual setup and disconnect by a network manager.

- Switched Virtual Connection (SVC) involves automatic setup and disconnection by a signaling protocol.
- An **ATM switch** is an electronic device with several input and output ports. It can process millions of cells per second.
- The ATM end user protocol consists of an **ATM adaptation layer**, an **ATM layer**, and a **Physical layer**.
- The ATM switch protocol consists of an ATM layer and Physical layer.
- **ATM Adaptation Layer Type1 (AAL1)** is designed to carry traffic with a constant bit rate.
- AAL3/4 is used to transfer variable length frames up to 65 k bytes.
- AAL5 is an efficient way to transfer large frames, such as LAN emulation.

Key Terms

ATM Adaptation Layer	Head of the Line Blocking
ATM Cell Format	Network-to-Network Interface (NNI)
ATM Forum	Permanent Virtual Connection (PVC)
ATM Layer	Switched Virtual Circuit (SVC)
ATM Protocol	User Network Interface (UNI)
Connection Identifiers	Virtual Channel Identifier (VCI)
Fabric Blocking	Virtual Path Identifier (VPI)

Review Questions

- **Multiple Choice Questions**

  1. ATM is a transfer protocol for _____.
     - a. ISDN
     - b. N-ISDN
     - c. B-ISDN
     - d. FDDI

  2. _____ is not a component of ATM.
     - a. Software
     - b. NIC
     - c. An ATM switch
     - d. A video card

  3. ATM has a fixed cell size of _____ bytes.
     - a. 10
     - b. 12
     - c. 53
     - d. 64

4. ATM cell has a _____-byte header.
   a. 1   c. 5
   b. 2   d. 10

5. ATM cell has a _____-byte payload.
   a. 1   c. 5
   b. 48   d. 28

6. ATM offers _____ types of connections.
   a. 6   c. 3
   b. 2   d. 1

7. An ATM switch will be involved with _____ type(s) of congestion.
   a. 1   c. 3
   b. 2   d. 4

8. ATM offers _____.
   a. bandwidth on demand   c. constant bandwidth
   b. variable bandwidth   d. none of the above

9. Which organization ratified the standard for ATM? _____
   a. ITU   c. EIA
   b. IEEE   d. IBM

10. ATM can be used for _____.
    a. LANs   c. LANs and WANs
    b. WANs   d. none of the above

11. The signaling protocol for ATM is used to set up a _____.
    a. permanent virtual connection   c. virtual connection
    b. switched virtual connection   d. none of the above

12. What type of ATM Adaptation layer is used for real-time video and voice transmission? _____
    a. AAL3/4   c. AAL1
    b. AAL2   d. AAL5

13. What type of AAL is used for transferring data efficiently? _____
    a. AAL2   c. AAL5
    b. AAL1   d. AAL3/4

14. ATM Class A service is used for _____.
    a. applications requiring a variable bit rate with connection-oriented
    b. applications requiring a constant bit rate with connection-oriented
    c. applications requiring a variable bit rate and connectionless
    d. none of the above

15. ATM class B service is used for _____.
    a. applications requiring a variable bit rate with connection-oriented
    b. applications requiring a constant bit rate with connection-oriented
    c. applications requiring a variable bit rate and connectionless
    d. none of the above

16. What type of AAL is used for Class A service? _____
    a. AAL2   c. AAL5
    b. AAL1   d. AAL3/4

17. What are the connection identifiers in an ATM cell header? _____
    a. VPI   c. VPI and VCI
    b. VCI   d. none of the above

18. The ATM protocol consists of _____.
    a. 3 layers   c. 1 layer
    b. 2 layers   d. 4 layers

19. An ATM switch consists of _____.
    a. 3 layers   c. 1 layer
    b. 2 layers   d. 4 layers

20. List the ATM layer architecture from top to bottom. _____
    a. ATM Adaptation layer, Physical layer, and ATM layer
    b. ATM Adaptation layer, ATM layer, and Physical layer
    c. ATM layer, ATM Adaptation layer, and Physical layer
    d. Physical layer, ATM Adaptation layer, and ATM layer

21. The function of ATM Adaptation is _____.
    a. segmentation and reassembly of the data unit
    b. cell header generation
    c. HEC generation
    d. transmit information to destination

22. The function of an ATM layer is _____.
    a. segmentation and reassembly of the data unit
    b. cell header generation
    c. HEC generation
    d. b and c

23. An ATM switch consists of the _____.
    a. ATM layer, ATM Adaptation layer, and Physical layer
    b. ATM layer and Physical layer
    c. ATM Adaptation layer and Physical layer
    d. Physical layer and ATM Adaptation layer

● **Short Answer Questions**

1. What does ATM stand for?

2. What is the data rate of ATM?

3. An ATM switch offers _____ types of interface?

4. What are the characteristics of ATM?

5. What are the advantages of ATM?

6. What is the data unit of ATM?

7. Explain cell switching.

8. Show an ATM cell.

9. How many byes are in an ATM payload?

10. How many bytes are in an ATM cell header?

11. How many bytes are in an ATM cell?

12. Explain Permanent Virtual Circuits.

13. Explain Switched Virtual Circuits.

14. What are connection identifiers?

15. What are the characteristics of an ATM switch?

16. What causes blocking in an ATM switch?

17. Show an ATM UNI cell header and explain the function of each field.

18. Show the ATM protocol.

19. Show an NNI ATM cell.

20. What is the function of the ATM Adaptation layer?

21. What is function of the ATM layer?

chapter 21

Network Operating System

OBJECTIVES

After completing this chapter, you should be able to:

- Describe the function of a network operating system
- Explain the function of BIOS and DOS
- Explain the function of NetBIOS
- List the most popular network operating systems
- Show the Windows NT networking model and explain the function of each component
- Show the Novell NetWare protocol architecture and describe the function of each protocol

INTRODUCTION

A Network Operating System (NOS) is an operating system that includes special functions for connecting computers and devices to local area networks, wide area networks, and the Internet. A network operating system enables users to communicate over networks and share resources such as files and printers. Some operating systems have networking functions built into them, such as UNIX, Linux, Windows NT, and Mac OS. Some popular network operating systems are Novell Netware, Windows NT, Banyan VINES, and Artisoft's LANtastic.

21.1 BIOS and DOS

The **Basic Input/Output System (BIOS)** is a set of control programs for input and output devices connected to the computer such as the video display, disk drives, printer and keyboard. These programs are stored in the ROM of the computer. The **Disk Operating System (DOS)** is a set of programs which provide a higher level interface with the BIOS. Figure 21.1 shows the relationship between DOS and BIOS. Non-Microsoft operating systems include functions similar to DOS.

FIGURE 21.1
BIOS and DOS
relation

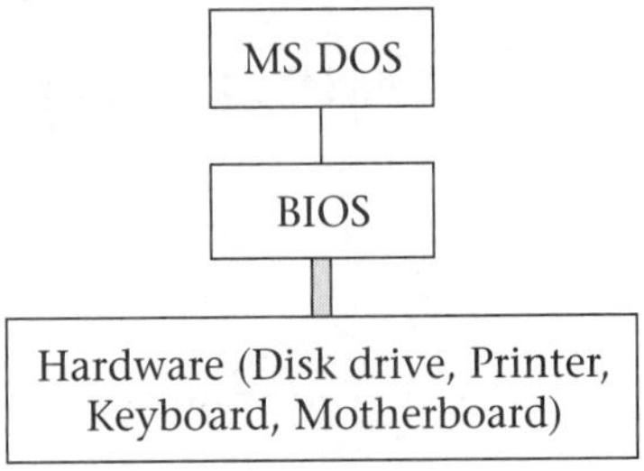

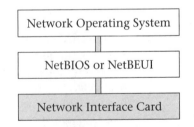

21.2 NetBIOS and NOS

The **Network Basic Input/Output System (NetBIOS)** is an interface specification developed by IBM as a standard for accessing a network. NetBIOS receives data from an upper layer and transmits the data to the network. An operating system that interfaces with NetBIOS is called a **Network Operating System (NOS)**. Computers are connected to a network through their NIC cards. The NIC card requires a special operating system to transmit information. This operating system is called NetBIOS and is stored in the ROM (read only memory) of the network interface card (NIC). The NetBIOS also provides a method for a network with multiple protocols to access networks. The NetBIOS is independent of the network hardware. Figure 21.2 shows the NetBIOS and NOS relationship.

FIGURE 21.2
Relation between
network operating
system and NetBIOS

21.3 Windows NT: New Technology

Windows NT is a 32-bit operating system designed for workstations and servers. Windows NT 3.5 and 4.0 support Pentium II, Pentium III, PowerPC, Alpha, and MIPS processors—by using a **Hardware Abstraction Layer (HAL)**. Window NT 4.0 is marketed in two versions: Windows NT for server and Window NT for workstation (client) or stand-alone. Windows NT supports a variety of protocols such as TCP/IP, IPX, NetBEUI, and AppleTalk. Figure 21.3 shows Windows NT protocols. Windows NT can operate in different network environments and supports interoperability with Novell NetWare and TCP/IP used on UNIX operating systems.

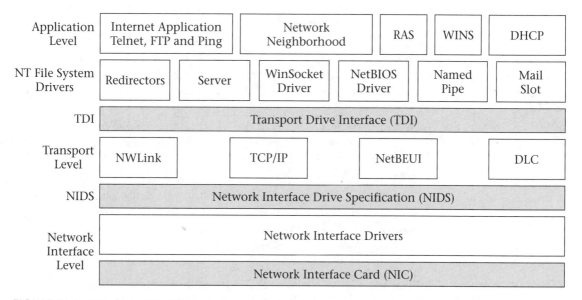

FIGURE 21.3 Windows NT communication model

The following paragraphs will describe the function of each level of the Windows NT communication model.

Application Level   The Application Level consists of the following:

Network Neighborhood: Network Neighborhood is a file manager which is used to view resources on the network.

Internet Application: Internet application consists of Telnet, FTP and Ping.

Remote Access Service (RAS): Windows NT uses RAS for connection to the remote host by modem or ISDN. RAS supports PPP and the obsolete SLIP protocol.

Window NT Name Service (WINS): The function of WINS is to hold a database which maps computer names to IP addresses.

Dynamic Host Configuration Protocol (DHCP): DHCP is used for dynamically assigning IP address to a station when the station wants to access the Internet.

A DHCP server stores all used and unused IP addresses of a network. The client requests an IP address and DHCP assigns an IP address to the client.

Windows NT File System Drivers

The following is a list of file system drivers for Windows NT:

Named Pipe: Named Pipe uses a connection-oriented mode to communicate between two applications. The source writes the information to a pipe (file) and a destination machine reads information from the pipe.

Mail Slot: Mail Slot uses a connectionless-oriented mode to transfer datagrams between two stations. Each station on the network can create a mail slot server and other clients on the network can store messages in the mail slot.

Server: The Server refers to the files that are needed for the redirector. The server receives a request from the redirector and accesses its file, then returns the result to the redirector.

Redirectors: The workstation receives the user request and passes the request to the redirector. A redirector passes the request for service to a server, then the server sends the response to the redirector.

Window Sockets: Window Sockets is an Application Program Interface that is used with UNIX operating systems.

Transport Driver Interface

The Transport Drive Interface (TDD) provides a general interface for drivers such as redirector, server, and Winsock to communicate with each transport protocol. TDI make all transport protocols look the same to the file system drivers.

Transport Level

The following is a list of file systems for the Transport level of Windows NT:

NetWare Link (NWLink): NWLink is a Microsoft Transport layer that is used for Internetworking. NWLink is a routable protocol and is an implementation of IPX/SPX (from Novell NetWare) for interoperability between Windows NT and Novell NetWare.

NetBEUI: Network BIOS Extended User Interface (NetBEUI) is an extended version of the NetBIOS protocol. NetBEUI is a transport protocol which provides services to a NetBIOS application. It is used by network operating systems such as Windows NT and LAN manager. NetBEUI is a small and fast protocol, which is used for small LANs with 20 to 200 users. NetBEUI cannot be used for routing information.

NetBIOS: In NetBIOS each computer on the network is assigned a name and NetBIOS uses these names to identify each computer. NetBIOS can use two types of connections: connectionless use for datagrams and connection-oriented use for service.

Data Link Control (DLC): The DLC protocol is used for accessing IBM mainframes and printers that are directly connected to a network.

Transmission Control Protocol and Internet Protocol (TCP/IP): TCP/IP is used for the Internet.

Network Interface Level

This level consists of the Network Device Interface Specifications (NDIS) and a Network Interface Card: NDIS is a Windows NT device driver interface that enables a single NIC to support multiple network protocols. NDIS can support both TCP/IP and IPX connections. NDIS accepts requests from the Transport layer and passes them to the Data Link layer. Novell NetWare calls NDIS an **Open Data Link Interface (ODI)**.

21.4 Novell NetWare Operating System

NetWare is a network operating system developed by Novell Inc. NetWare uses a client-server model architecture, in which a client requests service from a server. NetWare supports most computer operating systems such as DOS, Windows 95/98, UNIX, VMS and Macintosh. Figure 21.4 shows NetWare and the OSI Reference Model.

The following paragraphs describe the components of the Novell NetWare Operating System.

FIGURE 21.4
NetWare and OSI reference model

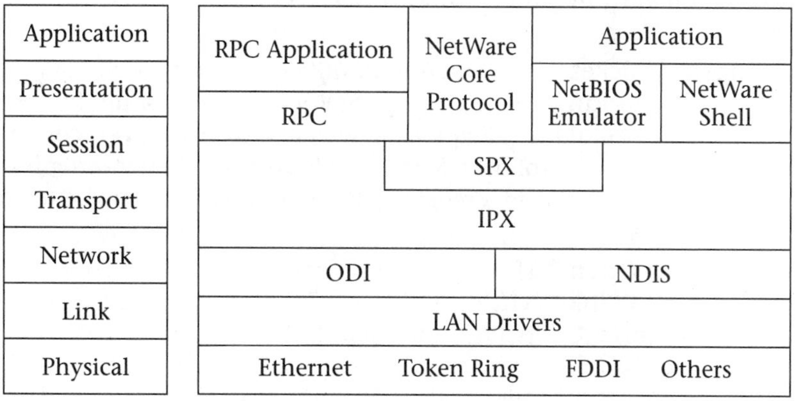

OSI	NetWare			
Application	RPC Application	NetWare Core Protocol	Application	
Presentation	RPC		NetBIOS Emulator	NetWare Shell
Session		SPX		
Transport	IPX			
Network	ODI		NDIS	
Link	LAN Drivers			
Physical	Ethernet	Token Ring	FDDI	Others

NetWare Core Protocol

NCP is a set of protocols run on a server to provide file service to client requests such as: File access, print service, security, connection, file locking, accounting, network management, and multiprotocol routing.

NetWare uses a modular architecture which can be expanded by adding NetWare loadable modules. Figure 21.5 shows the **NetWare Loadable Modules (NLM)** architecture.

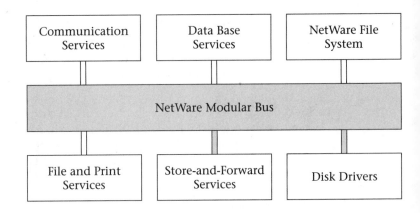

FIGURE 21.5
NetWare loadable modules

The following services are available through NLM: UNIX connectivity, host connectivity, message services, communication services, and network management.

NetWare Shell The NetWare shell runs on a workstation or client. It accepts a request from an application and determines if the request requires network access or not. If the request requires network access, the shell transmits the request to the lower layer for transmission to the server. If the application request does not require network access, the shell passes the application's request to the workstation operating system.

NetBIOS Emulation NetWare supports NetBIOS through emulation software for writing programs to interface with NetBIOS.

Internet Packet Exchange Protocol The function of IPX is to deliver packets to a destination. IPX performs packet routing using the Routing Information Protocol (RIP).
　　Sequenced Packet Exchange Protocol (SPX) operates at the Transport layer of the OSI model. SPX is a connection-oriented packet delivery protocol and provides reliable packet delivery from source to destination.

Network Open Data Link Interface The function of Network Open Data Link Interface (ODI) is to interface a single NIC or multiple NICs with different communication protocols such as IPX, TCP/IP, and AppleTalk.

Summary

- The **Networking Operating System** is an operating system which enable users to access networks and share resources such as files and printers.
- The **Basic Input/Output System (BIOS)** is set of programs stored in ROM. These programs are used to control input and output devices such as keyboards, disk drives, and printers.

- DOS is a set of programs from Microsoft which provide a higher level interface to the BIOS for the user.
- The most popular **Network Operating Systems** are UNIX, Linux, Windows NT, and Novell NetWare.
- The **Windows NT Application level** consists of: Internet Application (Telnet, FTP, and Ping), WINS, RAS, DHCP, and Network Neighborhood.
- The **Windows NT File System Drivers** are: Redirectors, Named Pipe, Window Sockets, Mail Slot, NetBIOS, and Server.
- The **Windows NT Transport level** consists of: NWLink, TCP/IP, NetBEUI, and DLC.
- The NetWare operating system is a network operating system developed by Novell Corp.
- Novell NetWare uses SPX/IPX as its Transport protocol.
- NetWare uses a set of protocols called the NetWare Core Protocol. NCP runs on the server and provides service to clients (stations).
- NetWare uses a modular architecture; each module can be loaded from a request.
- Client stations run the **NetWare Shell**. The function of the NetWare shell is to accept requests from users and pass the requests on to a server.

Key Terms

Basic Input/Output System (BIOS)

Disk Operating System (DOS)

Hardware Abstraction Layer (HAL)

NetWare Core Protocol

NetWare Link (NWLink)

NetWare Loadable Modules (NLM)

Network Basic Input/Output System (NetBIOS)

Network Operating System (NOS)

Open Data Link Interface (ODI)

Remote Access Service

Transport Drive Interface

Windows NT

Review Questions

- **Multiple Choice Questions**
    1. BIOS is a set of _____ programs.
        - a. small
        - b. large
        - c. control
        - d. none of the above

2. _____ is/are the operating system(s) designed for servers and workstations.
   a. DOS
   b. Windows NT
   c. NetBIOS
   d. UNIX and Windows NT

3. Windows NT uses _____ for connection to the remote host.
   a. FTP
   b. e-mail
   c. RAS
   d. Telnet

4. _____ is used for the Internet.
   a. TCP/IP
   b. NetBIOS
   c. X.25
   d. FTP

5. NCP is a set of _____.
   a. rules
   b. protocols
   c. a and b
   d. programs

6. The NetWare Shell runs on a _____.
   a. UNIX
   b. workstation or client
   c. server
   d. b and c

7. NOVELL uses _____ as transport protocol.
   a. SPX/IPX
   b. TCP/IP
   c. NWLink
   d. BIOS

8. Windows NT stands for _____.
   a. Windows Network
   b. Windows New Technology
   c. a and b
   d. Windows Network Technology

9. Windows NT is a _____-bit operating system.
   a. 16
   b. 32
   c. 64
   d. 8

● **Short Answer Questions**

1. What is the function of a Network Operating System?
2. Explain the function of the BIOS and its relation with DOS.
3. Explain the function of NetBIOS.
4. What does Windows NT stand for?
5. How many bits is the Windows NT operating system based on?

6. List the different types of Windows NT.
7. List the components of the Windows NT Application level.
8. What is the function of DHCP?
9. List the protocols that run on Windows NT.
10. What is NetWare?
11. List the NetWare core protocols.
12. What type of station runs NetWare Shell?
13. Name the transport protocol for Novell NetWare.

Computer and Communication Connectors

Data Connectors

Computer manufacturers use five different shell sizes for connectors—size 1 through size 5 (as shown in Figure A.1). Table A.1 shows the size of each shell. Each shell size offers two types of connector: standard DB and HD (high density). The DB connectors use two rows of pins and HD connectors use three rows of pins. HD-15 is used for video connectors.

FIGURE A.1
Five different shells

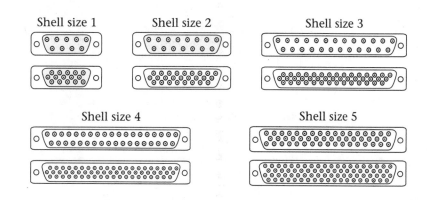

TABLE A.1 Shell Sizes

Shell Type	Size in Inches
Size 1	1.2 x 0.5
Size 2	1.55 x 0.5
Size 3	2.0 x 0.5
Size 4	2.7 x 0.5
Size 5	2.6 x 0.5

Serial Data Transmission

Serial transmission is one of the methods of transmitting data from one device to another device. Following are the serial interfaces used for serial transmission.

EIA RS-232 or
RS-232C Standard

The RS-232 interface defines the electrical function of the pins and the mechanical function of the connector.

The EIA revised RS-232C in 1989 and called the revision RS-232D (connector with 25 pins). RS-232 is a standard connection for serial communication, which has been approved by the EIA. All modems use RS-232 connections and all PCs have a RS-232 port. RS-232 supports two types of connectors: a 25-pin D-type connector (DB-25) as shown in Figure A.2 and a 9-pin D-type connector (DB-9) as shown in Figure A.3. The EIA has approved RS-422 and RS-423 which are successors to EIA RS-232.

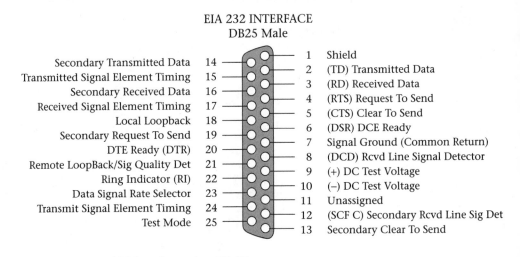

FIGURE A.2   EIA-232 interface using DB-25

FIGURE A.3
EIA-232 interface
using DB-9

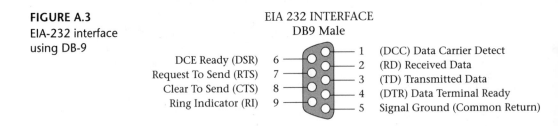

In RS-232, zero is represented by a voltage level of +3 volts to +12 volts and one is represented by –3 volts to –12 volts. RS-232 is used for 20 Kbps up to 50 feet.

RS-449 Interface
Standard

The most significant shortcoming of the RS-232 is its limited transmission rate and distance. The EIA approved RS-449 as an improved version of RS-232 using DB-37 conductors for serial transmission (as shown in Figure A.4). It is designed to increase the bandwidth and the distance of the cable. RS-449 is used to transmit data up to 10 Mbps and a distance of 40 feet or transmit data at a rate of 100 Kbps with 4000 feet maximum distance.

FIGURE A.4
RS-449 interface

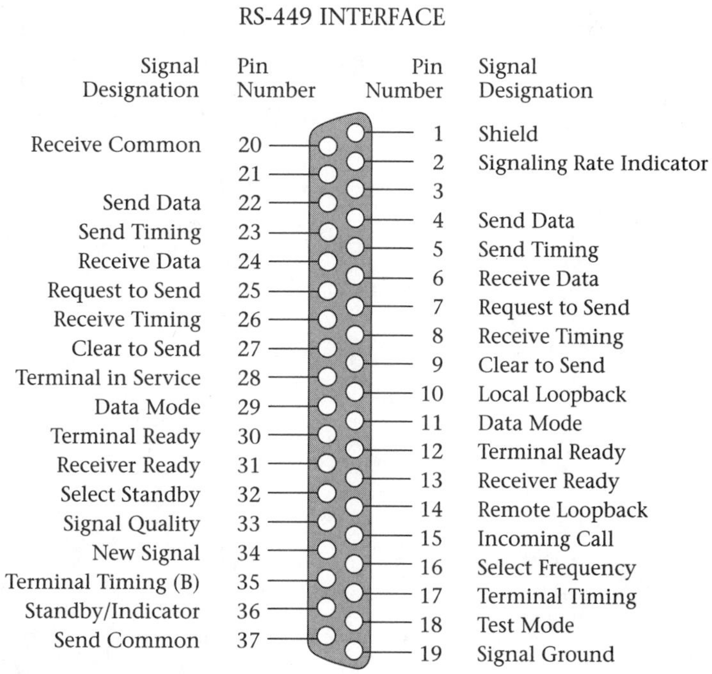

RS-449 INTERFACE

Signal Designation	Pin Number		Pin Number	Signal Designation
			1	Shield
Receive Common	20		2	Signaling Rate Indicator
	21		3	
Send Data	22		4	Send Data
Send Timing	23		5	Send Timing
Receive Data	24		6	Receive Data
Request to Send	25		7	Request to Send
Receive Timing	26		8	Receive Timing
Clear to Send	27		9	Clear to Send
Terminal in Service	28		10	Local Loopback
Data Mode	29		11	Data Mode
Terminal Ready	30		12	Terminal Ready
Receiver Ready	31		13	Receiver Ready
Select Standby	32		14	Remote Loopback
Signal Quality	33		15	Incoming Call
New Signal	34		16	Select Frequency
Terminal Timing (B)	35		17	Terminal Timing
Standby/Indicator	36		18	Test Mode
Send Common	37		19	Signal Ground

V.35 Standard The V.35 standard was developed by ITU for interfacing DTE or DEC to a high-speed digital carrier. The most common application of V.35 is for interfacing routers or DSU to a T-1 link. Figure A.5 shows the V.35 interface.

Null Modem The application of a null modem is to connect two PC serial ports together in order to transfer information between two PCs. Figure A.6 shows the connection between two RS-232 DB-9 connectors.

X.21 The X.21 interface is defined by ITU and uses balanced circuit transmission (two lines are used to transmit each signal; each line uses opposite polarity). The X.21 interface uses 15 pins. X.21 is used to connect DTE to public data networks such as ISDN.

Small Computer System Interface (SCSI)

The SCSI standard is defined by ANSI 9 American National Standard Institute (ANSI) for connecting daisy-chained, multiple I/O devices such as scanners, hard

FIGURE A.5
V.35 interface

V.35 INTERFACE

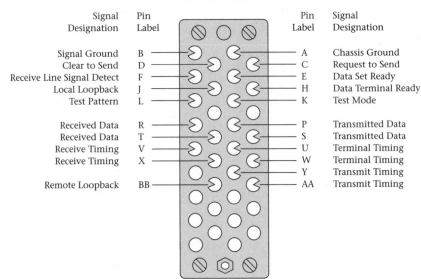

Signal Designation	Pin Label		Pin Label	Signal Designation
Signal Ground	B		A	Chassis Ground
Clear to Send	D		C	Request to Send
Receive Line Signal Detect	F		E	Data Set Ready
Local Loopback	J		H	Data Terminal Ready
Test Pattern	L		K	Test Mode
Received Data	R		P	Transmitted Data
Received Data	T		S	Transmitted Data
Receive Timing	V		U	Terminal Timing
Receive Timing	X		W	Terminal Timing
			Y	Transmit Timing
Remote Loopback	BB		AA	Transmit Timing

FIGURE A.6
Null modem
connection

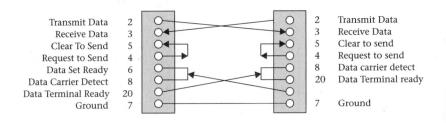

Transmit Data	2		2	Transmit Data
Receive Data	3		3	Receive Data
Clear To Send	5		5	Clear to send
Request to Send	4		4	Request to send
Data Set Ready	6		8	Data carrier detect
Data Carrier Detect	8		20	Data Terminal ready
Data Terminal Ready	20			
Ground	7		7	Ground

disks, and CD-ROMs to a microcomputer as shown in Figure A.7. Figure A.8 shows SCSI connectors.

FIGURE A.7
Several devices
connected by
daisy chaining

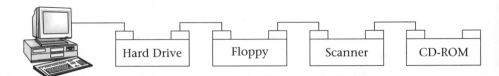

Parallel Connectors

Parallel transmission is used for high-speed transmission of data. Figure A.9 shows a DB-25 female parallel connector for a PC port. Parallel cable is limited to ten feet. IEEE-1284 enhanced parallel cable for high-speed printing is required. Figure A.10 shows Centronics connectors.

FIGURE A.8
SCSI-1, SCSI-2, SCSI-3, and SCSI-5 connectors

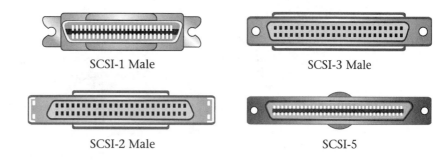

SCSI-1 Male

SCSI-3 Male

SCSI-2 Male

SCSI-5

FIGURE A.9
Parallel interface using DB-25

PARALLEL INTERFACE
DB25 Female

Return/Ground	25	13 Select (Active High)
Return/Ground	24	12 Paper End (Active High)
Return/Ground	23	11 Busy (Active LOW)
Return/Ground	22	10 Acknowledge (Active LOW)
Return/Ground	21	9 Data Bit 6 (MS8)
Return/Ground	20	8 Data Bit 7
Return/Ground	19	7 Data Bit 6
Return/Ground	18	6 Data Bit 5
Select Input (Active LOW)	17	5 Data Bit 4
Initialize Printer (Prime-Active LOW)	16	4 Data Bit 3
Error (Fault-Active LOW)	15	3 Data Bit 2
Auto Line Feed (Active LOW)	14	2 Data Bit 1 (LSB)
		1 DataStrobe (Active LOW)

Centronic female

Centronic male

FIGURE A.10 Centronics connectors

Modular Connectors

RJ-11, RJ-12, and RJ-45 are used with unshielded twisted-pair cable (UTP). RJ-45 uses 4 pairs of wires, RJ-12 uses 3 pairs of wires, and RJ-11 uses 2 pairs of wires. Figure A.11 shows these RJ connectors.

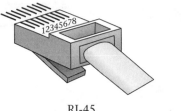 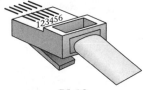 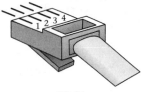

RJ-45   RJ-12   RJ-11

FIGURE A.11 RJ connectors

DIN-Type Connectors DIN connectors are used for keyboards and mice and come in four different types. Figure A.12 shows the DIN connectors listed below.

1. 4-Pin Mini DIN
2. 5-Pin Mini DIN (used for an AT-style keyboard)
3. 6-pin Mini DIN (used for PS/2-style keyboard)
4. 8-pin Mini DIN (used for Sun keyboards)

FIGURE A.12
DIN connectors

4-Pin Mini DIN   5-Pin DIN   6-Pin Mini DIN   8-Pin Mini DIN
(Female)   (Female)   (Female)   (Female)

Coaxial Cable Connectors:

There are three types of connectors listed below, which are used in coaxial cable. They are shown in Figure A.13.

1. BNC connector
2. F-type connector
3. RCA-type connector

FIGURE A.13
Coaxial Cable connectors

RCA   F   BNC

Fiber Optic Connectors

There are three type of connectors used for fiber-optic cable, as shown in Figure A.14. They are listed below.

1. ST connector
2. SC connector
3. MTR-J connector

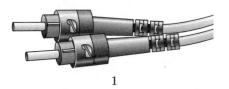

1

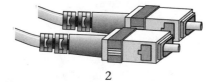

2

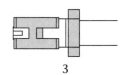

3
*MTRJ Connectors
*Dual Fiber/Single Jacket

FIGURE A.14 Fiber Optic connectors

FDDI Connector   Figure A.15 shows an FDDI connector.

FIGURE A.15
FDDI connector

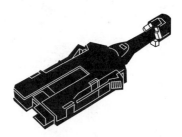

*FDDI (FSD) Connectors
*Ceramic Ferrules

Answers to Odd-Numbered Questions

CHAPTER 1

Multiple Choice Questions

1. (a) computer network
3. (a) client, server
5. (b) communication server
7. (b) ring
9. (c) Star
11. (b) 10
13. (a) Bus
15. (a) LAN

Short Answer Questions

1. A server is a station on a network that holds a Network Operating System and application software.
3. The client submits information to the server; the server processes the information and then transfers the result to the client.
5. NIC, Media, and NOS
7. Bus, Ring, Star, Fully-Connected, and Hybrid
9. A hub is a networking device that accepts information from one port and retransmits information to all ports.
11. Metropolitan Area Network
13. Wide Area Network

CHAPTER 2

Multiple Choice Questions

1. (a) $F = 1/T$
3. (a) 16
5. (a) Serial
7. (a) full-duplex
9. (a) 500 Kbps
11. (a) Manchester Encoding
13. (a) 1F7
15. (a) broadband signal
17. (a) 70%

Short Answer Questions

1. Voltage

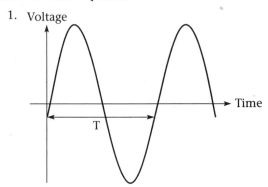

3. Hertz

5. The amplitude of an analog signal is its voltage at any given time.

7. One digit in binary is called a bit and is represented as a 0 or 1.

9. Two or more than two bytes are called a word.

11. $(11111111)_2 = 1 * 2^0 + 1 * 2^1 + 1 * 2^2 + 1 * 2^3 + 1 * 2^4 + 1 * 2^5 + 1 * 2^6 + 1 * 2^7 = 255$
    $(10110001)_2 = 181$

13.

1000100	1001001	1000111	1001001	1010100	1000001	1001100
D	I	G	I	T	A	L

15. ASCII to hexadecimal: Answers will vary.

17. In parallel transmission, multiple bits are transmitted simultaneously.

19. a. Simplex. The transmission of information goes in one direction.
    b. Half-Duplex. Two devices transmit information to each other, but one at a time.
    c. Full-Duplex. Both devices can receive and send information simultaneously.

21. A clock pulse is needed in order for the receiver to recognize the speed of incoming data.

23. Clock pulse is alternating ones and zeros.

Voltage

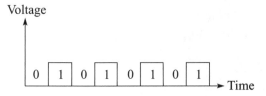

25. Parity bit, Block Check Character (BCC), Cyclic Redundancy Check (CRC), and One's complement of the sums

27. $X^5 + X^4 + X^2 + 1$

29.

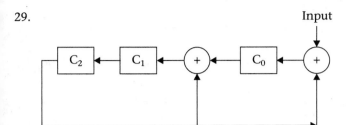

31.

Character	ASCII
N	1001110
E	1000101
T	1010100
W	1010111
O	1001111
R	1010010
K	1001011

0101010 and the one's complement is 1010101

33. Draw Manchester encoding and Differential Manchester encoding for binary 010110110

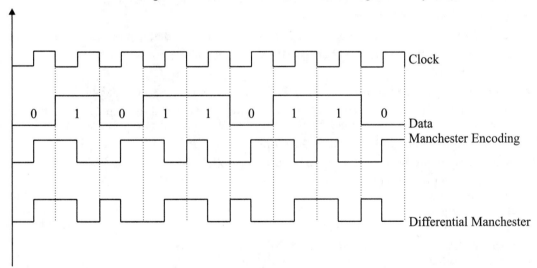

35. 200 − 120 = 80 KHz

37. There is a burst error when two more consecutive bits in a frame have changed.

CHAPTER 3

Multiple Choice Questions

1. (c) ALU
3. (b) reduced instruction set
5. (c) SDRAM and DRAM
7. (a) CPU
9. (b) PCI and EISA

Short Answer Questions

1. Microprocessor, memory, programmable interrupt, programmable parallel input/output, direct memory access, and bus
3. An Arithmetic Logic Unit perform arithmetic operations (addition, subtraction) and logic operations (AND, OR, and NOT).
5. The components of a CPU are the ALU, registers, and control unit. If these three units are in more than one integrated circuit, it is a CPU. If these three components are in one IC, it is a microprocessor.
7. Static Random Access Memory (SRAM) is used for cache memory.
9. The address bus carries the address of the information and the data bus carries the information.
11. $2^{10} * 8 = 1024 * 8$ bits or 1024 bytes
13. Electrically Erasable Read Only Memory (EEPROM) is a type of ROM that can be erased electrically.
15. Cache memory is fast memory located between the CPU and main memory.
17. Floppy disk, hard disk, and tape
19. Devices such as a printer and scanner with parallel ports are connected to the parallel port of the PC.
21. There are several peripherals connected to the PC. An interrupt is used to inform the CPU that a specific device does not work, for example, "printer out of paper" and "disk drive is not ready."
23. Windows 95, Windows NT, and Macintosh operating systems
25. Integrated Device Electronics, which is a type of disk controller
27. The characteristics of a Complex Instruction Set Computer (CISC) are:
    a. Control unit is microcode
    b. Uses long instruction set
    c. Uses indexed and indirect addressing

    The characteristics of a Reduced Instruction Set Computer (RISC) are:
    a. Control Unit is hardware
    b. Use small instruction sets
    c. Use a large number of registers

CHAPTER 4

Multiple Choice Questions

1. (a) IEEE 802
3. (c) ANSI
5. (c) NWLink
7. (a) Network
9. (b) Logical Link Control layer
11. (b) Presentation layer
13. (a) Physical layer
15. (c) IEEE 802.2
17. (b) Bit-oriented

Short Answer Questions

1. Developing standards for computers enables hardware products and software products to be compatible.
3. An open system is a set of protocols that allows two computers to communicate with each other, regardless of the manufacturer, CPU, and operating system.
5. The Application layer enables users to access networks through tools such as Telnet, FTP, and Network Neighborhood.
7. The function of the Session layer is to set up a session between two applications, such as "login" in Telnet.
9. Setting up a physical connection, disconnecting a connection, and routing the information are the functions of the Network layer.
11. The Physical layer provides electrical and mechanical interface with communications media and converts electrical and optical signals to bits, and vice versa.
13.

1 byte	1 byte	1 or 2 bytes	
DSAP	SSAP	Control	Information

15. IEEE802.3 is a standard for the MAC and Physical layers in Ethernet networks.
17. Asynchronous and Synchronous
19. Synchronous Data Link Control
21. High-Level Data Link Control defines the frame format for the Data Link layer.
23. 01111110 represents the start and end flag. If information field contains 01111110, the receiver will detect it as the end or start of the frame. To avoid this problem, the transmitter inserts an extra zero after five ones are repeated in the information field. The receiver will discard this extra zero.
25. Electronic Industry Association (EIA)
27. TCP/IP, NetBEUI, IPX/SPX, and NWLink
29. Data Link layer
31. Network Layer

33. Session layer

35.

SYN	SYN	STX		Information	ETX

37. SYN = 0010110, STX = 0000010, and ETX = 0000011
39. After bit insertion the result is: 0111110110000000011111100111110110

CHAPTER 5

Multiple Choice Questions

1. (a) UTP
3. (b) ST and SC
5. (c) electromagnetic
7. (a) Coaxial cable and fiber-optic cable

Short Answer Questions

1. Conductors, optical cable, and wireless communication
3. Shielded Twisted-Pair cable
5. CAT-5 UTP can handle signals with speeds up to 100 MHz.
7. a. Can transmit information in a longer distance
   b. Immune to the external noise
   c. Hard to tap the cable (more secure than conductor)
9. Single Mode Fiber cable is used for long distance transmission; only one ray of light can travel through it.
11. Long-distance communication
13. Terrestrial and satellite.
15. Laser Diode and Light Emitting Diode (LED)
17. In STP, the cable signal is more immune to noise.

CHAPTER 6

Multiple Choice Questions

1. (a) multiplexer
3. (b) SPM and FPM
5. (b) T1
7. (b) analog to digital
9. (b) 4 kHz

Short Answer Questions

1.

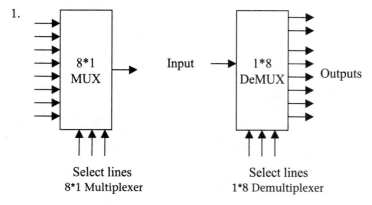

Select lines
8*1 Multiplexer

Select lines
1*8 Demultiplexer

3. In Time Division Multiplexer (TDM) operation, each input is assigned equal time to transmit its information.

5. Digital signal

7. PCM stands for Pulse Code Modulation. It is a method used to convert human voice to a digital signal.

9. Time Division Multiplexer

11. 1.54 Mbps

13. 44.736 Mbps

15. a. Circuit Switching: A physical connection must be established between the source and the destination before transmission of data begins.
   b. Message Switching: The entire message is transmitted to the switch. The switch stores the message, then retransmits the message to the next switch or destination.
   c. Packet Switching: The source station breaks the message into packets and transmits the packet to the switch; the switch stores the packets and retransmits them to the next switch or destination. In packet switching each packet might go via a different route in order to get to the destination.
   d. Virtual Circuit: All packets belong to a message and take a specific route to get to the destination.

17. One input, 16 outputs, and 4 select lines

19. A T1 link can carry 24 voice channels; the data rate for each channel is 64 Kbps. Therefore, the T1 link data rate = 24 * 64 Kbps.

21. The sampling rate is twice the highest frequency (2000 samples per second).

CHAPTER 7

Multiple Choice Questions

1. (c) a and b

3. (a) 33.6 Kbps

5. (b) DMT

7. (c) 160 Kbps

9. (b) DSL modem

11. (c) QPSK

13. (c) 6 Mhz

Short Answer Questions

1. Modulation and Demodulation

3. The data rate is defined by the number of bits per second.

5. Amplitude shift keying, or amplitude modulation, represents changes in the amplitude of the signal by ones and zeros.

7. Phase shift keying represents changes in the phase of the signal by binary ones and zeros.

9. 56 Kbps

11. Data rate is the number of bits per second and baud rate is the number of signals per second.

13. Digital Subscriber Line

15. ADSL uses current telephone wiring. These wires can handle signals up to 1 MHz. This bandwidth is divided into 4 kHz channels. The first channel is used for telephone conversation and 249 channels are used for transmitting and receiving information.

17. DSL is implemented in several different technologies called xDSL. The types of xDLS are: ADSL, HDSL, SDSL, and VDSL.

19. Yes. The data rate of the ADSL is dependent on the length of the cable between the subscriber and the telephone central switch.

21. Head end, trunk cable, feeder cable, amplifier, and drop cable

23. The bandwidth of a TV channel is 6 MHz.

25. 64-QAM (Qaudrature Amplitude Modulation) or 256-QAM

27. BaseT Repeater or PC with 10BaseT NIC

29. In FSK, each signal is represented by one bit; therefore, the data rate is 300 bps.

31. $2400/600 = 4$ bits

33. $2^5 = 32$, 25 Kbps/5 = 5,000 signals per second

35. Narrowband ISDN and Broadband ISDN

37. The B-channel is 64 Kbps and the D-channel is 16 Kbps.

39. Two telephones and one computer

CHAPTER 8

Multiple Choice Questions

1. (a) Ethernet

3. (a) CSMA/CD

5. (a) Manchester

7. (a) 10BaseT

9. (b) 185

11. (b) 10Base2

Short Answer Questions

1. a. 10BaseT: 10 Mbps, baseband, using twisted-pair cable as transmission media.
   b. 10Base5 or ThickNet: 10 Mbps, baseband, using thick coaxial cable for transmission media. The length of a network segment is 5×10 meters.
   c. 10Base2 or ThinNet: 10 Mbps, baseband, using thin coaxial cable as transmission media. The length of a network segment is 185 meters.

3. Unshielded Twisted-Pair cable and Shielded Twisted-Pair cable

5. The maximum length of the network without a repeater

7. A Repeater accepts information from one port and retransmits it to other ports.

9. Any station requiring access to the network performs the following tasks:
   a. Listens to the network. If the network is busy, the station keeps listening. If the network is not busy, it transmits information and listens for collision.
   b. Any station detecting collision will transmit a jam signal.

11. A standard for Logical Link Control (LLC)

13. Also called physical address, the MAC address is the 48 bits assigned to the NIC by its manufacturer.

15. When two stations transmit information on the Ethernet Network at the same time, a collision will occur.

17. A station uses broadcast address in the destination field of the frame to send the frame to all stations on the network. The destination address field for broadcast is set to all ones.

19. 48 bits

21. The function of a transceiver is to receive information from network media and transmit information to the network media

23. a. 400 meters
    b. 2000 meters
    c. 740 meters

25. 1512 bytes

27. From their hardware address (MAC address)

29. Error detection

31. To identify the number of bytes in an information field

33. LLC and MAC

CHAPTER 9

Multiple Choice Questions

1. (c) IEEE 802.5
3. (b) ring
5. (d) ring
7. (c) 8
9. (a) Token
11. (b) The station that turns on first

Short Answer Questions

1. The computers are connected to Multiple Access Units (MAUs) to form a ring connection.

3. 8 stations

5. a. 72 stations
   b. 260 stations

7. An active monitor performs the following functions
   a. Generates a token
   b. Remove unwanted frame from the ring
   c. Purge the ring in case of ring failure
   d. Broadcast a MAC frame every seven seconds to inform stations that there is an active monitor
   e. Detect lost of token and frames

9. a. Physical insertion of station
   b. Station sends MAC frames to the MAU to test connection
   c. The station will verify its unique address
   d. Find out address of its upstream neighbor

11. Differential Manchester Encoding

13. 24 bits

15. J and K bits are violations of Differential Manchester encoding (J-K bits do not convert to Differential Manchester encoding); therefore, the SD and ED fields do not appear in information field of the frame.

CHAPTER 10

Multiple Choice Questions

1. (c) IEEE 802.3u

3. (c) convergence sublayer

5. (a) 100

7. (b) 2

9. (a) 100

11. (a) 1

Short Answer Questions

1. a. 100Base4T: 100 Mbps, baseband, and using 4 pairs CAT-3 UTP as transmission media
   b. 100BaseTX: 100 Mbps, baseband, and using 2 pairs of UTP as transmission media
   c. 100BaseFX: 100 Mbps, baseband, and using fiber-optical cable as transmission media

3. 100BaseTX uses 2 pairs of CAT-5 UTP and 100Base4T uses 4 pairs of CAT-3 UTP.

5. A Class II repeater is used to connect stations with the same type of Fast Ethernet cards.

7. IEEE 802.3u

9. The convergence sublayer is used to interface the MAC layer with Physical layer.

11. 8B/6T (8 bits to 6 bits ternary)

CHAPTER 11

Multiple Choice Questions

1. (c) IEEE 802.12
3. (c) 4
5. (b) Promiscuous mode
7. (c) 4
9. (a) VVV

Short Answer Questions

1. Star topology
3. IEEE 802.12
5. UTP, STP, and fiber-optics cable
7. When the node is configured for promiscuous mode, the node receives all frames.
9. The maximum number of nodes connected to a 100 VG–AnyLAN repeater is 32.

11.

Start of Stream	Preamble	DA	SA	Request Config.	Allow Config.	Data	FCS	End of Stream

Start of Stream Delimiters (SSD)
  High Priority SSD: 0101111100000011
  Normal Priority SSD: 0101100000111110
End of Stream Delimiters (ESD)
  High Priority ESD: 111111 000011 000001
  Normal Priority ESD: 000000 111100 111110
Destination Address (DA): DA is 6 bytes
Requested Configuration Field Formats (2 bytes)
  The request configuration is two bytes and it is used by any nodes or repeaters to inform the higher-level repeater about requests for port configuration.
Allowed Configuration Field Format
  This is a response to a Request Configuration field by a higher-level repeater or node; the training initiator sets all bits of this field to zero.
Frame Check Sequence Field (FCS): IEEE uses a 32-bit cyclic redundancy check (CRC).

CHAPTER 12

Multiple Choice Questions

1. (c) switch
3. (b) asymmetric switching
5. (d) Network

7. (d) proxy
9. (a) Layer 2 switch
11. (b) Layer 3 switch
13. (a) Store-and-forward

Short Answer Questions

1. The switch is used to connect different segments of a LAN in order to increase network throughput.
3. The asymmetric switch provides switching between segments having different bandwidth: 10 Mbps to 100 Mbps or 100 Mbps to 1000 Mbps.
5. A store-and-forward switch stores the packet, then checks for errors. If there is no error in the packet, it retransmits the packet to the destination port. If the packet contains an error, it discards the packet
7. Data Link layer
9. Network layer
11. The L4 switch offers quality of service (QoS) to the packet.

CHAPTER 13

Multiple Choice Questions

1. (b) 1000
3. (a) CSMA/CD
5. (b) fiber-optic
7. (c) a and b
9. (a) CAT-5 UTP
11. (c) Both a and b

Short Answer Questions

1. IEEE 802.3z
3. Ethernet Frame
5. Multimode Fiber, Single Mode Fiber, UTP, and Shielded Balanced Twisted Pair
7. a. Gigabit Ethernet Interface Card
   b. Switches that can handle 100 Mbps Fast Ethernet and 1000 Mbps Ethernet
   c. Gigabit Ethernet Repeater
   d. Gigabit Switch

CHAPTER 14

Multiple Choice Questions

1. (d) repeater
3. (a) Bridges

5. (b) Routers
7. (b) transparent

Short Answer Questions

1. Repeaters, Bridges, Routers, Switches, Gateways
3. To expand the diameter of a network.
5. A bridge is used to connect different segments of a network.
7. Data Link layer
9. A source routing bridge routes the frame based on information in the routing field of the frame. Token ring frame format contains the routing field.
11. The routing table is set up by a network administrator.
13. Network layer
15. A dynamic router automatically generates a routing table.
17. A gateway is used to connect different network architectures such as DecNet and Ethernet.
19. A router is used to connect networks using IEEE 802.2 as logical link control. A gateway operates in the Application layer of the OSI model, while a router operates in Network layer of OSI model.

CHAPTER 15

Multiple Choice Questions

1. (a) Air
3. (b) CSMA/CA
5. (c) cell
7. (b) IEEE 802.11

Short Answer Questions

1. Wireless Local Area Network
3. An access point acts like a bridge, receiving information and retransmit information through the air and its wired port.
5. IEEE 802.11
7. Infrared and Radio Frequency
9. 1 watt
11. Industrial, Scientific, and Medical Band

CHAPTER 16

Multiple Choice Questions

1. (a) LAN
3. (a) 2

5. (c) 1000
7. (a) 2
9. (b) campus backbone

Short Answer Questions

1. Fiber Distributed Data Interface.
3. 100 Mbps
5. An FDDI concentrator is used as the ring and the stations are connected to the concentrator.
7. SAS is a Single Attachment Station, which attaches to only primary ring of FDDI.
9. The optical bypass switch is used to isolate stations from the ring.
11. 4B/5B (4-bit binary to 5-bit symbol)
13. Fast Ethernet NIC is less expensive than FDDI NIC. Fast Ethernet can use UTP as a transmission media, which is less expensive than fiber cable.

CHAPTER 17

Multiple Choice Questions

1. (c) ITU
3. (a) STS-1
5. (a) regenerator
7. (c) Optical Carrier
9. (b) PC-1
11. (b) three STS-3s
13. (b) a frame with a data rate less than STS-1

Short Answer Questions

1. Synchronous Optical Network
3. SONET is a high-speed optical data carrier.
5. Fiber-optic cable.
7. The SONET components are:
   a. STS multiplexer
   b. STS demultiplexer
   c. Add/Drop multiplexer
   d. Regenerator
9. Optical Carrier type 1 (OC-1). SONET convert STS-1 to OC-1 (Optical Signal) with the data rate of 51.84 Mbps.
11. 810 bytes
13. Three

15. A VT2 frame is 36 bytes transmitted at 8000 frames per second; therefore, the data rate is $36 \times 8 \times 8000$.

17. Indicates the start of frame inside of the SPE

19. Synchronous Transport Signal Type n

21. a. Payload Pointer: Specifies the location of the SPE.
    b. Path Overhead: A performance monitor of the STS, path trace, parity check and path status.
    c. Line Overhead: Carries the Payload Pointers and provides automatic switching. Also separates voice channels, provides multiplexing and line maintenance.
    d. Section Overhead: Contains information about frame synchronization, carries information about OAM, handles frame alignment, and separates data from voice.

CHAPTER 18

Multiple Choice Questions

1. (c) ISDN
3. (b) at the central office of a telephone company
5. (b) WAN
7. (c) FRAD
9. (a) the LANs that are located in different cities

Short Answer Questions

1. Both the Data Link layer and Physical layer of the OSI model
3. The function of FRAD (Frame Relay Assembler and Disassembler) is to convert the frame format from a site network to the frame format of the frame relay and vice versa.
5. a. Frame relay access equipment: Customer premises equipment, such as a LAN and a Frame Relay Assembler and Disassembler (FRAD), such as a router
   b. Frame relay switches
   c. Frame Relay Service provided by: public carriers

CHAPTER 19

Multiple Choice Questions

1. (c) networks
3. (d) Antivirus software
5. (a) TCP
7. (c) IETF
9. (b) UDP
11. (c) PPP
13. (b) 32 bits
15. (a) 191.205.205.255
17. (a) reliable communication

19. (a) The source reports its buffer size to the destination.
21. (b) TCP
23. (a) IP address
25. (a) An IP subnet mask represents bits in the network portion of AN IP address.

Short Answer Questions

1. HTML (Hypertext Markup Language) is used to create files for the WWW (World Wide Web).
3. Point-to-Point Protocol (PPP) or SLIP
5. TCP (Transmission Control Protocol) and UDP (User Datagram Protocol)
7. 32 bits
9. The function of IP is to deliver packets to a destination.
11. 20 bytes
13. 8 bytes
15. $2^8 = 256$
17. Requesting physical addresses from a destination.
19. IPv4
21. 128 bits
23. E-mail, Telnet, WWW, and FTP
25. TFTP (Trivial File Transfer Protocol) and RCP (Remote Call Procedure)
27. The University of California at Berkeley
29. Internet Control Message Protocols is used for handling error messages and control messages.
31. Telnet is one of the application protocols of TCP/IP and is used for remote connection of one computer to another computer.
33. When a packet reaches the TCP, TCP must transfer the packet to the Application layer. TCP uses a port number to identify the application protocol.
35. a. Class B
    b. Class D
    c. Class A
37. a. 7 bits
    b. 512
    c. 255.254.0.0

CHAPTER 20

Multiple Choice Questions

1. (c) B-ISDN
3. (c) 53
5. (b) 48
7. (b) 2
9. (a) ITU

11. (a) permanent virtual connection
13. (c) AAL5
15. (c) applications requiring a variable bit rate and connectionless
17. (c) VPI and VCI
19. (b) 2 layers
21. (a) segmentation and reassembly of the data unit
23. (b) ATM layer and Physical layer

Short Answer Questions

1. Asynchronous Transfer Mode
3. Two
5. Permanent Virtual circuit and Switched Virtual Circuit
7. a. Offers bandwidth on the demand
   b. Can transfer many types of information such as voice, video, and data
   c. Supports variable and constant bit rates
9. An ATM switch offers cell switching, meaning it reads the incoming cell header and finds the proper route to transmit the cell without storing the cell.
11. 48 bytes
13. 53 bytes
15. A set of network switches between ATM source and destination are programmed with predefined values for VCI/VPI. PVC is used for more reliable transmission.
17. ATM switches must process million of cells per second, modify the VPI/VCI field of an incoming cell, transmit the cell to the proper output port, and check the header for errors.
19. Fabric blocking and Head of the line blocking
21. The ATM Adaptation layer converts the large Service Data Unit (SDU) to a 48-byte ATM cell payload. AAL is designed in several types in order to transfer all types of information.

CHAPTER 21

Multiple Choice Questions

1. (a) small
3. (c) RAS
5. (b) protocols
7. (a) SPX/IPX
9. (b) 32

Short Answer Questions

1. A Network Operating System (NOS) is an operating system that enables a user to access a network for the purpose of sharing files and printers.
3. Network Basic Input/Output System (NetBIOS) is a set of programs that receive data from an upper layer and retransmit it to the network.

5. Windows NT is a 32-bit operating system.

7. Dynamic Host Configuration Protocol (DHCP) is used to assign an IP address to a station in a network dynamically.

9. NetWare is a Network Operating System by Novell Corp.

11. Workstation and client

13. Internet Packet Exchange Protocol (IPX) and Sequenced Packet Protocol (SPX)

Acronyms

A

AAL   ATM Adaptation Layer

ACK   Acknowledgment

ADSL   Asymmetrical Digital Subscriber Line

ALU   Arithmetic Logic Unit

AM   Amplitude Modulation

AMI   Alternate Mark Inversion

ANSI   American National Standards Institute

API   Application Program Interface

ARP   Address Resolution Protocol

ARPA   Advanced Research Project Agency

ARPANET   Advanced Research Project Agency Network

ASCII   American Standard Code for Information Interchange

ASIC   Application Specific Integrated Circuit

ASK   Amplitude Shift Keying

ATM   Asynchronous Transfer Mode

AUI   Attachment Unit Interface

B

B Channel   Bearer Channel

BCC   Block Check Character

BCD   Binary Coded Decimal

B-ISDN   Broadband Integrated Services Digital Network

BBN   Bolt Beranek & Newman

BENC   Backward Explicit Congestion Notification

BIOS   Basic Input/Output System

BNC   British Naval Connector

BRI   Basic Rate Interface

BSD   Berkeley Software Distribution

C

CAD   Computer-Aided Design

CAT   Categories

CBR Constant Bit Rate

CD-ROM Compact Disk-Read Only Memory

CISC Complex Instruction Set Computer

CLP Congestion Loss Priority

CPI Common Port Identification

CPU Central Processing Unit

C/R Command/Response

CRC Cyclic Redundancy Check

CS Convergence Sublayer

CSMA/CD Carrier-Sense Multiple-Access/Collision Detection

CSU/DSU Channel Service Unit/Data Service Unit

D

DA Destination Address

DARPA Defense Advanced Research Projects Agency

DAS Dual Attachment Station

DE Discard Eligibility

DF Do Not Fragment

DEMUX Demultiplexer

DIMM Dual In-Line Memory Module

DLC Data Link Control

DLCI Data Link Connection Identifier

DMA Direct Memory Access

DMT Discrete Multi-Tone

DNS Domain Name System

DOD Department of Defense

DOS Disk Operating System

DPAM Demand Priority Access Method

DRAM Dynamic Random Access Memory

DSAP Destination Service Access Point

DSL Digital Subscriber Line

DS-n Digital Signal Level n

DTE Data Terminal Equipment

E

EEPROM Electrically Erasable Programmable Read Only Memory

EHF Extreme High Frequency

EIA   Electronics Industries Association
EIDE   Extended Integrated Disk Electronics
EISA   Extension Industry Standard Architecture
EPROM   Erasable Programmable Read Only Memory
ESC   Escape
ESD   End of Stream Delimiters
ETX   End of Text

F

FCC   Federal Communication Commission
FCS   Frame Check Sequence
FDDI   Fiber Distributed Data Interface
FDM   Frequency Division Multiplexing
FECN   Forward Explicit Congestion Notification
FM   Frequency Modulation
FRAD   Frame Relay Assembler and Disassembler
FSK   Frequency Shift Keying
FTP   File Transfer Protocol

G

GFC   Generic Flow Control

H

HDLC   High-Level Data Link Control
HDSL   High-Bit-Rate Digital Subscriber Line
HEC   Header Error Control
HELN   Hardware Address Length
HFC   Hybrid Fiber Coax
HTML   Hypertext Markup Language
HTTP   Hypertext Transfer Protocol

I

IAB   Internet Architecture Board
IANA   Internet Assigned Numbers Authority
IBM   International Business Machines
IC   Integrated Circuit
ICMP   Internet Control Message Protocol

IDE   Integrated Disk Electronics
IEEE   Institute of Electrical and Electronic Engineers
IESG   Internet Engineering Steering Group
IETF   Internet Engineering Task Force
IF   Information Field
INIC   Internet Network Information Center
I/O   Input/Output
IP   Internet Protocol
IPv6   Internet Protocol Version 6
IPX/SPX   Internet Packet Exchange/Sequence Packet Exchange
IRTF   Internet Research Task Force
ISA   Industry Standard Architecture
ISDN   Integrated Services Digital Network
ISO   International Organization for Standardization
ITU   International Telecommunications Union

L

LAA   Locally Administered Address
LAN   Local Area Network
LAP   Link Access Protocol
LAPB   Link Access Procedure Balanced
LED   Light-Emitting Diode
LD   Laser Diode
LLC   Logical Link Control
L3 Switch   Layer 3 Switch
LW   Long Wave

M

MAC   Media Access Control
MAN   Metropolitan Area Network
MAU   Multiple Access Unit or Multistation Access Unit
MCA   Micro-Channel Architecture
MF   More Fragment
MII   Media Independent Interface
MMF   Multimode Fiber
MTU   Maximum Transfer Unit
MUX   Multiplexer
MW   Medium Wave

N

NCP   Network Control Protocol

NDIS   Network Device Interface Specification

NetBEUI   NetBIOS Extended User Interface

NetBIOS   Network Basic Input/Output System

NFS   Network File System

NIC   Network Interface Card

NISDN   Narrowband Integrated Services Digital Network

NLM   Network Loadable Modules

NNI   Network-to-Network Interface

NOS   Network Operating System

NRZ   Non-Return to Zero

NRZ-I   Non-Return to Zero Inverted

NRZ-L   Non-Return to Zero Level

NSF   National Science Foundation

NT   New Technology

NT1   Network Termination Device (Type 1)

O

OAM   Operation, Administration and Maintenance

OC   Optical Carrier

ODI   Open Data Link Interface

OSI   Open System Interconnection

P

PAM   Pulse Amplitude Modulation

PC   Personal Computer

PCI   Peripheral Component Interconnection

PCM   Pulse Code Modulation

PCMCIA   Personal Computer Memory Card International Association

PDM   Physical Medium Dependent

PDU   Protocol Data Unit

POTS   Plain Old Telephone Service

PnP   Plug-and-Play

PPP   Point-to-Point Protocol

PRI   Primary Rate Interface

PSK   Phase Shift Keying

PSTN   Packet Switched Telephone Network

PT Payload Type
PVC Permanent Virtual Connection
PVC Polyvinyl Chloride

Q

QAM Quadrature Amplitude Modulation

R

RADSL Rate Adaptive Asymmetric Digital Subscriber Line
RAM Random Access Memory
RARP Reverse Address Resolution Protocol
RG Radio Government
RI Routing Information
RISC Reduced Instruction Set Computer
ROM Read Only Memory
RPC Remote Call Procedure
RZ Return to Zero

S

SA Source Address
SAR Segmentation and Reassembly
SAS Single Attachment Station
SCSI Small Computer Systems Interface
SDH Synchronous Digital Hierarchy
SDLC Synchronous Data Link Control
SDRAM Synchronous DRAM
SDSL Symmetrical Digital Subscriber Line
SFD Start Frame Delimiter
SHF Super High Frequency
SIE Serial Engine Interface
SIMD Single Instruction Multiple Data
SIMM Single In-Line Memory Module
SMF Single Mode Fiber
SMTP Simple Mail Transfer Protocol
SMT Station Management
SNA System Network Architecture
SNMP Simple Network Management Protocol
SONET Synchronous Optical Network

SPE   Synchronous Payload Envelope
SPID   Service Profile Identifiers
SPM   Statistical Packet Multiplexing
SQE   Signal Quality Error
SRAM   Static Random Access Memory
SSAP   Source Service Access Point
STM   Synchronous Transport Module
STP   Shielded Twisted Pair
STS   Synchronous Transport Signal
STX   Start of Text
SVC   Switched Virtual Connection
SW   Short Wave

T

TA   Terminal Adapter
TC   Transmission Convergence
TCP   Transmission Control Protocol
TCP/IP   Transmission Control Protocol/Internet Protocol
TDM   Time Division Multiplexing
TFTP   Trivial File Transfer Protocol
TOS   Type Of Service
TTL   Time To Live

U

UAA   Universal Administered Address
UDP   User Datagram Protocol
UHF   Ultra High Frequency
UNI   User Network Interface
URL   Uniform Resource Locator
UTP   Unshielded Twisted Pair

V

VBR   Variable Bit Rate
VCI   Virtual Channel Identifier
VDSL   Very High-Speed Digital Subscriber Line
VESA   Video Electronics Standards Association
VHF   Very High Frequency
VLAN   Virtual Local Area Network

VLF   Very Low Frequency
VPI   Virtual Path Identifier
VRC   Vertical Redundancy Check

W

WAN   Wide Area Network
Windows NT   Windows New Technology
WWW   World Wide Web

Index

O

100BaseFX media, for Fast
    Ethernet, 150, 151–153, 156
100BaseT4 media, for Fast Ethernet,
    150, 151
100BaseTX media, for Fast Ethernet,
    150, 151, 152, 154–155, 156
100 Mbps Voice Grade-Any Local
    Area Network. *See* 100
    VG–AnyLAN technology
100 VG–AnyLAN technology,
    159–165
  described, 159–160
  end node operation with,
    160–161
  frame format with, 163–165
  frames with, 162
  repeaters with, 161
  transmission media with, 162
One's complement of the sum, in
    signal error detection, 29,
    30–31
1000BaseCX transmission medium,
    for Gigabit Ethernet, 183,
    184, 185, 186
1000BaseLX transmission medium,
    for Gigabit Ethernet, 183,
    184, 185, 186
1000BaseSX transmission medium,
    for Gigabit Ethernet, 183,
    184, 185, 186
1000BaseT transmission medium,
    for Gigabit Ethernet, 183,
    184, 185, 186
Open System Interconnection (OSI)
    model, 61–73
  with FDDI technology, 214–216
  IEEE 802 Standard for, 70–73
  standardization organizations
    and, 60
Operation Administration and
    Maintenance (OAM), with
    SONET, 224
Overheads, in SONET frames,
    225–226

P

Packet filtering, as firewall, 175
Packets. *See also* Datagrams; Frames

with ARP, 255–256
with frame relay, 237–238
in Internet, 248–249
with IPv6, 263–267
with PPP, 260–261
with TCP, 250–251
Packet switching, 96, 97
Pad Field, in Ethernet frames, 126
Parallel connectors, 47
  SCSI, 49–50, 308–309
Parallel input/output interfaces,
    41–42, 47
Parallel ports, with USB, 53
Parallel transmission, 23, 24
  with USB, 53
Parity bit
  in asynchronous transmission,
    22–23
  with block check character,
    29–30
  cyclic redundancy check and,
    31
  with one's complement of the
    sum, 30–31
  in parity check, 29
Parity-bit generator, 29, 30
Parity check, in signal error
    detection, 29
Path overhead, in SONET frames,
    226, 227, 228
Payload Length field, in IPv6
    packets, 263, 264
Payload Type (PT) field, in ATM
    cells, 281–282
PC cards, 52
PCI cards, 51–52
Peer-to-peer networks, 2
  WLANs as, 201–202
Pentium processors, 42–43, 54–55
Period, of sine wave, 15–16
Periodic analog signals, 16–17
Peripheral component interconnect
    (PCI) buses, 51–52
Peripherals, with USB, 53
Permanent virtual circuits (PVCs),
    with frame relay, 237
Permanent Virtual Connection
    (PVC), with ATM, 277
Phase, of analog signal, 16–17,
    103–105, 106
Phase Shift Keying (PSK), 102,
    103–105

Physical Interface Layer, for Gigabit
    Ethernet, 183–184
Physical layer
  with ATM, 282–284
  for Ethernet, 124
  with FDDI technology, 214
  for frame relay, 234
  for Gigabit Ethernet, 185
  IEEE 802 Standard for, 70–73
  of OSI model, 61, 62, 194
  for WLANs, 205–208
Physical Medium Dependent
    (PMD) sublayer
  with ATM, 282–284
  for Fast Ethernet, 150, 151
  with FDDI technology, 214,
    215
Physical Sublayer Protocol (PHY),
    with FDDI technology, 215
Plain Old Telephone Service
    (POTS), 108
  with ASDLs, 108–110
PLEN IP Address field, in ARP
    packets, 255, 256
Plug-and-play (PnP) technology, 52
Point-to-point links, HDLC and, 63
Point-to-Point Protocol (PPP), for
    Internet, 260–261
Polar encoding, 26–27
Port numbers, with TCP/IP, 251,
    252
Ports, with FDDI technology,
    216–217. *See also* Parallel
    ports; Serial ports; Uplink
    ports
Positive Acknowledgment Frame
    (ACK[N]), 67–68, 69
POTS filters, with ASDLs, 109
Preamble (PA)
  in Ethernet frames, 125
  in FDDI frames, 219
Precedence, of datagrams, 253
Presentation layer, of OSI model,
    62, 65
Primary Rate Interface (PRI),
    115–117
Primary ring, with FDDI
    technology, 214, 215,
    216–217, 218
Primary stations, with HDLC,
    62–63
Print servers, 3